SQA

Official

SQA Past Papers

WITH ANSWERS

Advanced Higher
Mathematics

2010–2014

Hodder Gibson Study Skills Advice – General — page 3
Hodder Gibson Study Skills Advice –
 Advanced Higher Mathematics — page 5
2010 EXAM — page 7
2011 EXAM — page 15
2012 EXAM — page 23
2013 EXAM — page 29
2014 EXAM — page 35
ANSWER SECTION — page 43

HODDER
GIBSON
AN HACHETTE UK COMPANY

Hodder Gibson is grateful to the copyright holders, as credited on the final page of the Question Section, for permission to use their material. Every effort has been made to trace the copyright holders and to obtain their permission for the use of copyright material. Hodder Gibson will be happy to receive information allowing us to rectify any error or omission in future editions.

Hachette UK's policy is to use papers that are natural, renewable and recyclable products and made from wood grown in sustainable forests. The logging and manufacturing processes are expected to conform to the environmental regulations of the country of origin.

Orders: please contact Bookpoint Ltd, 130 Park Drive, Abingdon, Oxon OX14 4SE. Telephone: (44) 01235 827720. Fax: (44) 01235 400454.

Lines are open 9.00–5.00, Monday to Saturday, with a 24-hour message answering service. Visit our website at www.hoddereducation.co.uk. Hodder Gibson can be contacted direct on: Tel: 0141 848 1609; Fax: 0141 889 6315; email: hoddergibson@hodder.co.uk

This collection first published in 2014 by

Hodder Gibson, an imprint of Hodder Education,

An Hachette UK Company

2a Christie Street

Paisley PA1 1NB

{BrightRED Hodder Gibson is grateful to Bright Red Publishing Ltd for collaborative work in preparation of this book and all SQA Past Paper, National 5 and CfE Higher Model Paper titles 2014.

Typeset by PDQ Digital Media Solutions Ltd, Bungay, Suffolk NR35 1BY

Printed in the UK

A catalogue record for this title is available from the British Library

ISBN 978-1-4718-3674-9

3 2 1

2015 2014

Introduction

Study Skills – what you need to know to pass exams!

Pause for thought

Many students might skip quickly through a page like this. After all, we all know how to revise. Do you really though?

Think about this:

"IF YOU ALWAYS DO WHAT YOU ALWAYS DO, YOU WILL ALWAYS GET WHAT YOU HAVE ALWAYS GOT."

Do you like the grades you get? Do you want to do better? If you get full marks in your assessment, then that's great! Change nothing! This section is just to help you get that little bit better than you already are.

There are two main parts to the advice on offer here. The first part highlights fairly obvious things but which are also very important. The second part makes suggestions about revision that you might not have thought about but which WILL help you.

Part 1

DOH! It's so obvious but …

Start revising in good time

Don't leave it until the last minute – this will make you panic.

Make a revision timetable that sets out work time AND play time.

Sleep and eat!

Obvious really, and very helpful. Avoid arguments or stressful things too – even games that wind you up. You need to be fit, awake and focused!

Know your place!

Make sure you know exactly **WHEN and WHERE** your exams are.

Know your enemy!

Make sure you know what to expect in the exam.

How is the paper structured?

How much time is there for each question?

What types of question are involved?

Which topics seem to come up time and time again?

Which topics are your strongest and which are your weakest?

Are all topics compulsory or are there choices?

Learn by DOING!

There is no substitute for past papers and practice papers – they are simply essential! Tackling this collection of papers and answers is exactly the right thing to be doing as your exams approach.

Part 2

People learn in different ways. Some like low light, some bright. Some like early morning, some like evening / night. Some prefer warm, some prefer cold. But everyone uses their BRAIN and the brain works when it is active. Passive learning – sitting gazing at notes – is the most INEFFICIENT way to learn anything. Below you will find tips and ideas for making your revision more effective and maybe even more enjoyable. What follows gets your brain active, and active learning works!

Activity 1 – Stop and review

Step 1

When you have done no more than 5 minutes of revision reading STOP!

Step 2

Write a heading in your own words which sums up the topic you have been revising.

Step 3

Write a summary of what you have revised in no more than two sentences. Don't fool yourself by saying, "I know it, but I cannot put it into words". That just means you don't know it well enough. If you cannot write your summary, revise that section again, knowing that you must write a summary at the end of it. Many of you will have notebooks full of blue/black ink writing. Many of the pages will not be especially attractive or memorable so try to liven them up a bit with colour as you are reviewing and rewriting. **This is a great memory aid, and memory is the most important thing.**

Activity 2 — Use technology!

Why should everything be written down? Have you thought about "mental" maps, diagrams, cartoons and colour to help you learn? And rather than write down notes, why not record your revision material?

What about having a text message revision session with friends? Keep in touch with them to find out how and what they are revising and share ideas and questions.

Why not make a video diary where you tell the camera what you are doing, what you think you have learned and what you still have to do? No one has to see or hear it, but the process of having to organise your thoughts in a formal way to explain something is a very important learning practice.

Be sure to make use of electronic files. You could begin to summarise your class notes. Your typing might be slow, but it will get faster and the typed notes will be easier to read than the scribbles in your class notes. Try to add different fonts and colours to make your work stand out. You can easily Google relevant pictures, cartoons and diagrams which you can copy and paste to make your work more attractive and **MEMORABLE**.

Activity 3 – This is it. Do this and you will know lots!

Step 1

In this task you must be very honest with yourself! Find the SQA syllabus for your subject (www.sqa.org.uk). Look at how it is broken down into main topics called MANDATORY knowledge. That means stuff you MUST know.

Step 2

BEFORE you do ANY revision on this topic, write a list of everything that you already know about the subject. It might be quite a long list but you only need to write it once. It shows you all the information that is already in your long-term memory so you know what parts you do not need to revise!

Step 3

Pick a chapter or section from your book or revision notes. Choose a fairly large section or a whole chapter to get the most out of this activity.

With a buddy, use Skype, Facetime, Twitter or any other communication you have, to play the game "If this is the answer, what is the question?". For example, if you are revising Geography and the answer you provide is "meander", your buddy would have to make up a question like "What is the word that describes a feature of a river where it flows slowly and bends often from side to side?".

Make up 10 "answers" based on the content of the chapter or section you are using. Give this to your buddy to solve while you solve theirs.

Step 4

Construct a wordsearch of at least 10 X 10 squares. You can make it as big as you like but keep it realistic. Work together with a group of friends. Many apps allow you to make wordsearch puzzles online. The words and phrases can go in any direction and phrases can be split. Your puzzle must only contain facts linked to the topic you are revising. Your task is to find 10 bits of information to hide in your puzzle, but you must not repeat information that you used in Step 3. DO NOT show where the words are. Fill up empty squares with random letters. Remember to keep a note of where your answers are hidden but do not show your friends. When you have a complete puzzle, exchange it with a friend to solve each other's puzzle.

Step 5

Now make up 10 questions (not "answers" this time) based on the same chapter used in the previous two tasks. Again, you must find NEW information that you have not yet used. Now it's getting hard to find that new information! Again, give your questions to a friend to answer.

Step 6

As you have been doing the puzzles, your brain has been actively searching for new information. Now write a NEW LIST that contains only the new information you have discovered when doing the puzzles. Your new list is the one to look at repeatedly for short bursts over the next few days. Try to remember more and more of it without looking at it. After a few days, you should be able to add words from your second list to your first list as you increase the information in your long-term memory.

FINALLY! Be inspired...

Make a list of different revision ideas and beside each one write **THINGS I HAVE** tried, **THINGS I WILL** try and **THINGS I MIGHT** try. Don't be scared of trying something new.

And remember – "FAIL TO PREPARE AND PREPARE TO FAIL!"

Advanced Higher Mathematics

The course

The Advanced Higher Mathematics Course emphasises the need for candidates to undertake extended thinking and decision making, to solve problems and integrate mathematical knowledge. The course offers candidates an enhanced awareness of the range and power of mathematics in an interesting and enjoyable manner.

As with all mathematics courses, Advanced Higher Mathematics aims to build upon and extend candidates' mathematical skills, knowledge and understanding in a way that recognises problem solving as an essential skill and enables them to integrate their knowledge of different aspects of the subject. The syllabus covers algebra, geometry and calculus. It includes matrix algebra, complex numbers, vectors and formalising the concept of mathematical proof.

How the course is graded

The grade you finally get for Advanced Higher Mathematics depends on two things:

- the internal assessments you do in school or college (the "NABs") – these don't count towards the final grade, but you must have passed them before you can achieve a final grade
- the examination you sit in May – that's what this book is all about!

The examination

The external examination, in which the use of a calculator is permitted, is a three hour paper with a total of 100 marks. It consists of short response questions, designed to test knowledge and understanding, as well as extended response questions to test problem solving skills. Following on from approximately twelve shorter style questions, there are four long questions totalling 40 marks.

The examination will test the candidate's ability to solve problems which integrate mathematical knowledge and which require extended thinking and decision making. The award of grades A, B and C is determined by the candidate's demonstration of the ability to apply knowledge and understanding to problem solving. To achieve grades A and B in particular, this demonstration will involve more complex contexts and include more in-depth treatment of content.

In solving these problems, candidates should be able to:

(a) interpret the problem and consider what might be relevant;

(b) decide how to proceed by selecting an appropriate strategy;

(c) implement the strategy through applying mathematical knowledge and understanding to come to a conclusion;

(d) decide on the most appropriate way of communicating the solution to the problem in an intelligible form.

Familiarity and complexity affect the level of difficulty of problems. It is generally easier to interpret and communicate information in contexts where the relevant variables are obvious and where their inter-relationships are known. It is usually more straightforward to apply a known strategy than to modify one or devise a new one. Some concepts are harder to grasp and some techniques more difficult to apply if they have to be used in combination.

The papers are designed so that approximately 65% of the marks will be opportunities at grade C.

The SQA gives detailed advice in the Candidate Guidance Information section of the Advanced Higher Mathematics page on its website http://www.sqa.org.uk/sqa/39095.html.

Preparation and hints

Remember to make connections between parts of questions, particularly where there are three or four sections to a question. These are almost always linked and, in some instances, an earlier result in part (a) or (b) is needed and its use would avoid further, repeated work. Where the word "Hence…" starts a later part of a question, candidates are expected to use their result(s) from an earlier part.

Below are some key tips for your success.

Accuracy

Avoid errors with minus signs, be careful when subtracting one expression from another and ensure that any negative is applied correctly. Do not make a straightforward question more difficult, or impossible, by allowing careless errors, particularly in calculus and algebra. Where possible use exact values as decimal approximations could cost marks.

Basic skills

Regular practice of the basic rules, such as the product and quotient rules, plus learning essential formulae will enable you to carry out routine steps automatically and without error. This will leave you time to concentrate on applying and integrating your skills and knowledge in more complex problems.

Communication

Communication is an important aspect of this examination and you are encouraged to show all working. This is particularly important in questions that contain the words "show that" or "prove". In this type of question you must get to the result quoted in the question; failure to do so will usually prevent you gaining the final mark, at least for that question or part question. The work leading to the result must be shown fully for marks to be awarded. Looking at such questions and the corresponding marking instructions is invaluable.

Higher knowledge

This course extends your mathematical knowledge and skills from Higher. You will need to know trigonometric identities and other relevant formulae.

Marking instructions

Ensure that you look at the detailed marking instructions of past papers. They provide further advice and guidelines as well as showing you precisely where, and for what, marks are awarded.

Notation

In all questions make sure that you use the correct notation. In particular, for logarithmic expressions, use modulus signs where necessary and, in many calculus questions, the constant of integration is essential.

Radians

Remember to work in radian measure when attempting any question involving both trigonometry and calculus.

Simplify

Get into the habit of simplifying expressions before doing any further work with them. This should make all subsequent work easier.

Working

You are encouraged to include all the relevant working, communicate answers clearly and write legibly; this will increase the opportunity to be awarded all of the marks possible in each question. Sometimes, there are many valid methods for a question and you will be able to access partial marks for appropriate working.

Good luck!

Remember that the rewards for passing Advanced Higher Mathematics are well worth it! Your pass will help you get the future you want for yourself. In the exam, be confident in your own ability. If you're not sure how to answer a question, trust your instincts and just give it a go anyway. Keep calm and don't panic! GOOD LUCK!

ADVANCED HIGHER

2010

[BLANK PAGE]

X100/701

NATIONAL
QUALIFICATIONS
2010

FRIDAY, 21 MAY
1.00 PM – 4.00 PM

MATHEMATICS
ADVANCED HIGHER

Read carefully

1. Calculators may be used in this paper.

2. Candidates should answer **all** questions.

3. **Full credit will be given only where the solution contains appropriate working.**

Marks

Answer all the questions.

1. Differentiate the following functions.

   (a) $f(x) = e^x \sin x^2$. 3

   (b) $g(x) = \dfrac{x^3}{(1 + \tan x)}$. 3

2. The second and third terms of a geometric series are −6 and 3 respectively. Explain why the series has a sum to infinity, and obtain this sum. 5

3. (a) Use the substitution $t = x^4$ to obtain $\displaystyle\int \frac{x^3}{1 + x^8}\, dx$. 3

   (b) Integrate $x^2 \ln x$ with respect to x. 4

4. Obtain the 2×2 matrix M associated with an enlargement, scale factor 2, followed by a clockwise rotation of $60°$ about the origin. 4

5. Show that

$$\binom{n+1}{3} - \binom{n}{3} = \binom{n}{2}$$

   where the integer n is greater than or equal to 3. 4

6. Given $\mathbf{u} = -2\mathbf{i} + 5\mathbf{k}$, $\mathbf{v} = 3\mathbf{i} + 2\mathbf{j} - \mathbf{k}$ and $\mathbf{w} = -\mathbf{i} + \mathbf{j} + 4\mathbf{k}$.

   Calculate $\mathbf{u}.(\mathbf{v} \times \mathbf{w})$. 4

Marks

7. Evaluate

$$\int_1^2 \frac{3x+5}{(x+1)(x+2)(x+3)}\, dx$$

expressing your answer in the form $\ln\frac{a}{b}$, where a and b are integers.   **6**

8. (*a*) Prove that the product of two odd integers is odd.   **2**

(*b*) Let p be an odd integer. Use the result of (*a*) to prove by induction that p^n is odd for all positive integers n.   **4**

9. Obtain the first three non-zero terms in the Maclaurin expansion of $(1 + \sin^2 x)$.   **4**

10. The diagram below shows part of the graph of a function $f(x)$. State whether $f(x)$ is odd, even or neither. Fully justify your answer.

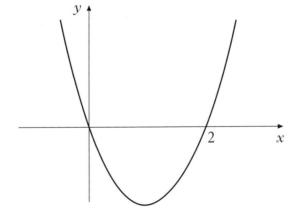

3

11. Obtain the general solution of the equation

$$\frac{d^2 y}{dx^2} + 4\frac{dy}{dx} + 5y = 0.$$ **4**

Hence obtain the solution for which $y = 3$ when $x = 0$ and $y = e^{-\pi}$ when $x = \frac{\pi}{2}$.   **3**

Marks

12. Prove by contradiction that if x is an irrational number, then $2 + x$ is irrational.   **4**

13. Given $y = t^3 - \dfrac{5}{2}t^2$ and $x = \sqrt{t}$ for $t > 0$, use parametric differentiation to express $\dfrac{dy}{dx}$ in terms of t in simplified form.   **4**

Show that $\dfrac{d^2y}{dx^2} = at^2 + bt$, determining the values of the constants a and b.   **3**

Obtain an equation for the tangent to the curve which passes through the point of inflexion.   **3**

14. Use Gaussian elimination to show that the set of equations

$$x - y + z = 1$$
$$x + y + 2z = 0$$
$$2x - y + az = 2$$

has a unique solution when $a \neq 2 \cdot 5$.   **5**

Explain what happens when $a = 2 \cdot 5$.   **1**

Obtain the solution when $a = 3$.   **1**

Given $A = \begin{pmatrix} 5 & 2 & -3 \\ 1 & 1 & -1 \\ -3 & -1 & 2 \end{pmatrix}$ and $B = \begin{pmatrix} 1 \\ 0 \\ 2 \end{pmatrix}$, calculate AB.   **1**

Hence, or otherwise, state the relationship between A and the matrix

$$C = \begin{pmatrix} 1 & -1 & 1 \\ 1 & 1 & 2 \\ 2 & -1 & 3 \end{pmatrix}.$$ **2**

Marks

15. A new board game has been invented and the symmetrical design on the board is made from four identical "petal" shapes. One of these petals is the region enclosed between the curves $y = x^2$ and $y^2 = 8x$ as shown shaded in diagram 1 below.

Calculate the area of the complete design, as shown in diagram 2.   **5**

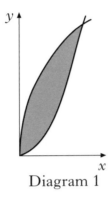

Diagram 1

Diagram 2

The counter used in the game is formed by rotating the shaded area shown in diagram 1 above, through $360°$ about the y-axis. Find the volume of plastic required to make one counter.   **5**

16. Given $z = r(\cos\theta + i\sin\theta)$, use de Moivre's theorem to express z^3 in polar form.   **1**

Hence obtain $\left(\cos\frac{2\pi}{3} + i\sin\frac{2\pi}{3}\right)^3$ in the form $a + ib$.   **2**

Hence, or otherwise, obtain the roots of the equation $z^3 = 8$ in Cartesian form.   **4**

Denoting the roots of $z^3 = 8$ by z_1, z_2, z_3:

(*a*) state the value $z_1 + z_2 + z_3$;

(*b*) obtain the value of $z_1^6 + z_2^6 + z_3^6$.   **3**

[END OF QUESTION PAPER]

[BLANK PAGE]

[BLANK PAGE]

X100/701

NATIONAL	WEDNESDAY, 18 MAY	MATHEMATICS
QUALIFICATIONS	1.00 PM – 4.00 PM	ADVANCED HIGHER
2011		

Read carefully

1. Calculators may be used in this paper.

2. Candidates should answer **all** questions.

3. **Full credit will be given only where the solution contains appropriate working.**

Marks

Answer all the questions.

1. Express $\dfrac{13-x}{x^2+4x-5}$ in partial fractions and hence obtain

$$\int \dfrac{13-x}{x^2+4x-5}dx.$$ 　　　5

2. Use the binomial theorem to expand $\left(\dfrac{1}{2}x-3\right)^4$ and simplify your answer. 　　　3

3. (a) Obtain $\dfrac{dy}{dx}$ when y is defined as a function of x by the equation

$$y+e^y=x^2.$$ 　　　3

　(b) Given $f(x)=\sin x\cos^3 x$, obtain $f'(x)$. 　　　3

4. (a) For what value of λ is $\begin{pmatrix} 1 & 2 & -1 \\ 3 & 0 & 2 \\ -1 & \lambda & 6 \end{pmatrix}$ singular? 　　　3

　(b) For $A=\begin{pmatrix} 2 & 2\alpha-\beta & -1 \\ 3\alpha+2\beta & 4 & 3 \\ -1 & 3 & 2 \end{pmatrix}$, obtain values of α and β such that

$$A'=\begin{pmatrix} 2 & -5 & -1 \\ -1 & 4 & 3 \\ -1 & 3 & 2 \end{pmatrix}.$$ 　　　3

5. Obtain the first four terms in the Maclaurin series of $\sqrt{1+x}$, and hence write down the first four terms in the Maclaurin series of $\sqrt{1+x^2}$. 　　　4

Hence obtain the first four terms in the Maclaurin series of $\sqrt{(1+x)(1+x^2)}$. 　　　2

Marks

6.

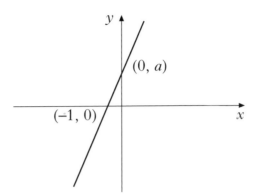

The diagram shows part of the graph of a function $f(x)$. Sketch the graph of $|f^{-1}(x)|$ showing the points of intersection with the axes.

4

7. A curve is defined by the equation $y = \dfrac{e^{\sin x}(2+x)^3}{\sqrt{1-x}}$ for $x < 1$.

Calculate the gradient of the curve when $x = 0$.

4

8. Write down an expression for $\displaystyle\sum_{r=1}^{n} r^3 - \left(\sum_{r=1}^{n} r\right)^2$

1

and an expression for

$$\sum_{r=1}^{n} r^3 + \left(\sum_{r=1}^{n} r\right)^2.$$

3

9. Given that $y > -1$ and $x > -1$, obtain the general solution of the differential equation

$$\frac{dy}{dx} = 3(1+y)\sqrt{1+x}$$

expressing your answer in the form $y = f(x)$.

5

[Turn over

Marks

10. Identify the locus in the complex plane given by

$$|z - 1| = 3.$$

Show in a diagram the region given by $|z - 1| \leq 3$.   **5**

11. (a) Obtain the exact value of $\int_0^{\pi/4} (\sec x - x)(\sec x + x)dx$.   **3**

    (b) Find $\int \dfrac{x}{\sqrt{1 - 49x^4}}dx$.   **4**

12. Prove by induction that $8^n + 3^{n-2}$ is divisible by 5 **for all integers $n \geq 2$**.   **5**

13. The first three terms of an arithmetic sequence are $a, \dfrac{1}{a}, 1$ where $a < 0$.

Obtain the value of a and the common difference.   **5**

Obtain the smallest value of n for which the sum of the first n terms is greater than 1000.   **4**

14. Find the general solution of the differential equation

$$\frac{d^2y}{dx^2} - \frac{dy}{dx} - 2y = e^x + 12.$$ **7**

Find the particular solution for which $y = -\dfrac{3}{2}$ and $\dfrac{dy}{dx} = \dfrac{1}{2}$ when $x = 0$.   **3**

Marks

15. The lines L_1 and L_2 are given by the equations

$$\frac{x-1}{k} = \frac{y}{-1} = \frac{z+3}{1} \quad \text{and} \quad \frac{x-4}{1} = \frac{y+3}{1} = \frac{z+3}{2},$$

respectively.

Find:

(*a*) the value of k for which L_1 and L_2 intersect and the point of intersection; **6**

(*b*) the acute angle between L_1 and L_2. **4**

16. Define $I_n = \int_0^1 \frac{1}{(1+x^2)^n} dx$ for $n \geq 1$.

(*a*) Use integration by parts to show that

$$I_n = \frac{1}{2^n} + 2n \int_0^1 \frac{x^2}{(1+x^2)^{n+1}} dx.$$

3

(*b*) Find the values of A and B for which

$$\frac{A}{(1+x^2)^n} + \frac{B}{(1+x^2)^{n+1}} = \frac{x^2}{(1+x^2)^{n+1}}$$

and hence show that

$$I_{n+1} = \frac{1}{n \times 2^{n+1}} + \left(\frac{2n-1}{2n}\right) I_n.$$

5

(*c*) Hence obtain the exact value of $\int_0^1 \frac{1}{(1+x^2)^3} dx.$ **3**

[*END OF QUESTION PAPER*]

[BLANK PAGE]

ADVANCED HIGHER

2012

[BLANK PAGE]

X100/13/01

NATIONAL
QUALIFICATIONS
2012

MONDAY, 21 MAY
1.00 PM – 4.00 PM

MATHEMATICS
ADVANCED HIGHER

Read carefully

1   Calculators may be used in this paper.

2   Candidates should answer **all** questions.

3   **Full credit will be given only where the solution contains appropriate working.**

Mark

Answer all the questions.

1. (a) Given $f(x) = \dfrac{3x+1}{x^2+1}$, obtain $f'(x)$.   **3**

    (b) Let $g(x) = \cos^2 x \exp(\tan x)$. Obtain an expression for $g'(x)$ and simplify your answer.   **4**

2. The first and fourth terms of a geometric series are 2048 and 256 respectively. Calculate the value of the common ratio.   **2**

    Given that the sum of the first n terms is 4088, find the value of n.   **3**

3. Given that $(-1 + 2i)$ is a root of the equation
$$z^3 + 5z^2 + 11z + 15 = 0,$$
obtain all the roots.   **4**

    Plot all the roots on an Argand diagram.   **2**

4. Write down and simplify the general term in the expansion of $\left(2x - \dfrac{1}{x^2}\right)^9$.   **3**

    Hence, or otherwise, obtain the term independent of x.   **2**

5. Obtain an equation for the plane passing through the points $P(-2, 1, -1)$, $Q(1, 2, 3)$ and $R(3, 0, 1)$.   **5**

6. Write down the Maclaurin expansion of e^x as far as the term in x^3.   **1**

    Hence, or otherwise, obtain the Maclaurin expansion of $(1 + e^x)^2$ as far as the term in x^3.   **4**

7. A function is defined by $f(x) = |x + 2|$ for all x.

    (a) Sketch the graph of the function for $-3 \le x \le 3$.   **2**

    (b) On a separate diagram, sketch the graph of $f'(x)$.   **2**

8. Use the substitution $x = 4 \sin \theta$ to evaluate $\displaystyle\int_0^2 \sqrt{16 - x^2}\ dx$.   **6**

Marks

9. A non-singular $n \times n$ matrix A satisfies the equation $A + A^{-1} = I$, where I is the $n \times n$ identity matrix. Show that $A^3 = kI$ and state the value of k.   **4**

10. Use the division algorithm to express 1234_{10} in base 7.   **3**

11. (a) Write down the derivative of $\sin^{-1}x$.   **1**

    (b) Use integration by parts to obtain $\displaystyle\int \sin^{-1}x . \frac{x}{\sqrt{1-x^2}}\,dx$.   **4**

12. The radius of a cylindrical column of liquid is decreasing at the rate of $0\cdot02$ m s^{-1}, while the height is increasing at the rate of $0\cdot01$ m s^{-1}.

    Find the rate of change of the volume when the radius is $0\cdot6$ metres and the height is 2 metres.   **5**

    [*Recall that the volume of a cylinder is given by $V = \pi r^2 h$.*]

13. A curve is defined parametrically, for all t, by the equations

    $$x = 2t + \frac{1}{2}t^2, \qquad y = \frac{1}{3}t^3 - 3t.$$

    Obtain $\dfrac{dy}{dx}$ and $\dfrac{d^2y}{dx^2}$ as functions of t.   **5**

    Find the values of t at which the curve has stationary points and determine their nature.   **3**

    Show that the curve has exactly two points of inflexion.   **2**

14. (a) Use Gaussian elimination to obtain the solution of the following system of equations in terms of the parameter λ.

    $$4x + 6z = 1$$
    $$2x - 2y + 4z = -1$$
    $$-x + y + \lambda z = 2$$

    5

    (b) Describe what happens when $\lambda = -2$.   **1**

    (c) When $\lambda = -1\cdot9$ the solution is $x = -22\cdot25$, $y = 8\cdot25$, $z = 15$.

       Find the solution when $\lambda = -2\cdot1$.   **2**

       Comment on these solutions.   **1**

    [Turn over for Questions 15 and 16 on *Page four*

Mark

15. (*a*) Express $\dfrac{1}{(x-1)(x+2)^2}$ in partial fractions.   **4**

(*b*) Obtain the general solution of the differential equation

$$(x-1)\frac{dy}{dx} - y = \frac{x-1}{(x+2)^2},$$

expressing your answer in the form $y = f(x)$.   **7**

16. (*a*) Prove by induction that

$$(\cos\theta + i\sin\theta)^n = \cos n\theta + i\sin n\theta$$

for all integers $n \geq 1$.   **6**

(*b*) Show that the real part of $\dfrac{\left(\cos\dfrac{\pi}{18} + i\sin\dfrac{\pi}{18}\right)^{11}}{\left(\cos\dfrac{\pi}{36} + i\sin\dfrac{\pi}{36}\right)^4}$ is zero.   **4**

[END OF QUESTION PAPER]

ADVANCED HIGHER
2013

[BLANK PAGE]

X100/13/01

NATIONAL
QUALIFICATIONS
2013

WEDNESDAY, 22 MAY
1.00 PM – 4.00 PM

MATHEMATICS
ADVANCED HIGHER

Read carefully

1 Calculators may be used in this paper.

2 Candidates should answer **all** questions.

3 **Full credit will be given only where the solution contains appropriate working.**

Mark

Answer all the questions.

1. Write down the binomial expansion of $\left(3x - \dfrac{2}{x^2}\right)^4$ and simplify your answer.   **4**

2. Differentiate $f(x) = e^{\cos x}\sin^2 x$.   **3**

3. Matrices A and B are defined by $A = \begin{pmatrix} 4 & p \\ -2 & 1 \end{pmatrix}$ and $B = \begin{pmatrix} x & -6 \\ 1 & 3 \end{pmatrix}$.

   (a) Find A^2.   **1**

   (b) Find the value of p for which A^2 is singular.   **2**

   (c) Find the values of p and x if $B = 3A'$.   **2**

4. The velocity, v, of a particle P at time t is given by
$$v = e^{3t} + 2e^t.$$

   (a) Find the acceleration of P at time t.   **2**

   (b) Find the distance covered by P between $t = 0$ and $t = \ln 3$.   **3**

5. Use the Euclidean algorithm to obtain the greatest common divisor of 1204 and 833, expressing it in the form $1204a + 833b$, where a and b are integers.   **4**

6. Integrate $\dfrac{\sec^2 3x}{1 + \tan 3x}$ with respect to x.   **4**

7. Given that $z = 1 - \sqrt{3}\,i$, write down $\bar{z}$ and express $\bar{z}^2$ in polar form.   **4**

8. Use integration by parts to obtain $\int x^2 \cos 3x\, dx$.   **5**

9. Prove by induction that, for all positive integers n,
$$\sum_{r=1}^{n} \left(4r^3 + 3r^2 + r\right) = n(n+1)^3.$$
   6

Marks

10. Describe the loci in the complex plane given by:

(a) $|z + i| = 1$;   **2**

(b) $|z - 1| = |z + 5|$.   **3**

11. A curve has equation

$$x^2 + 4xy + y^2 + 11 = 0.$$

Find the values of $\dfrac{dy}{dx}$ and $\dfrac{d^2y}{dx^2}$ at the point $(-2, 3)$.   **6**

12. Let n be a natural number.
For each of the following statements, decide whether it is true or false.
If true, give a proof; if false, give a counterexample.

A   If n is a multiple of 9 then so is n^2.

B   If n^2 is a multiple of 9 then so is n.   **4**

13. Part of the straight line graph of a function $f(x)$ is shown.

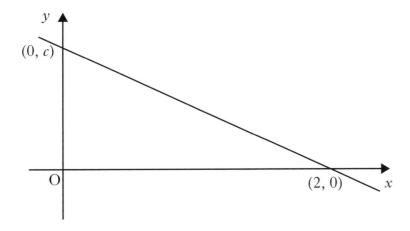

(a) Sketch the graph of $f^{-1}(x)$, showing points of intersection with the axes.   **2**

(b) State the value of k for which $f(x) + k$ is an odd function.   **1**

(c) Find the value of h for which $|f(x + h)|$ is an even function.   **2**

[Turn over for Questions 14 to 17 on *Page four*

Marks

14. Solve the differential equation

$$\frac{d^2 y}{dx^2} - 6\frac{dy}{dx} + 9y = 4e^{3x},$$ given that $y = 1$ and $\frac{dy}{dx} = -1$ when $x = 0.$ 11

15. (a) Find an equation of the plane π_1, through the points $A(0, -1, 3)$, $B(1, 0, 3)$ and $C(0, 0, 5)$.   4

   (b) π_2 is the plane through A with normal in the direction $-\mathbf{j} + \mathbf{k}$.

   Find an equation of the plane π_2.   2

   (c) Determine the acute angle between planes π_1 and π_2.   3

16. In an environment without enough resources to support a population greater than 1000, the population $P(t)$ at time t is governed by Verhurst's law

$$\frac{dP}{dt} = P(1000 - P).$$

   Show that

$$\ln \frac{P}{1000 - P} = 1000t + C \quad \text{for some constant } C.$$ 4

   Hence show that

$$P(t) = \frac{1000K}{K + e^{-1000t}} \quad \text{for some constant } K.$$ 3

   Given that $P(0) = 200$, determine at what time t, $P(t) = 900$.   3

17. Write down the sums to infinity of the geometric series

$$1 + x + x^2 + x^3 + \ldots\ldots$$

   and

$$1 - x + x^2 - x^3 + \ldots\ldots$$

   valid for $|x| < 1$.

   Assuming that it is permitted to integrate an infinite series term by term, show that, for $|x| < 1$,

$$\ln\left(\frac{1+x}{1-x}\right) = 2\left(x + \frac{x^3}{3} + \frac{x^5}{5} + \ldots\ldots\right).$$ 7

   Show how this series can be used to evaluate $\ln 2$.

   Hence determine the value of $\ln 2$ correct to 3 decimal places.   3

[END OF QUESTION PAPER]

ADVANCED HIGHER

2014

[BLANK PAGE]

X100/13/01

NATIONAL QUALIFICATIONS 2014	TUESDAY, 6 MAY 1.00 PM – 4.00 PM	MATHEMATICS ADVANCED HIGHER

Read carefully

1   Calculators may be used in this paper.

2   Candidates should answer **all** questions.

3   **Full credit will be given only where the solution contains appropriate working.**

Mark

Answer all the questions.

1. (a) Given

$$f(x) = \frac{x^2 - 1}{x^2 + 1},$$

obtain $f'(x)$ and simplify your answer.   **3**

(b) Differentiate $y = \tan^{-1}(3x^2)$.   **3**

2. Write down and simplify the general term in the expression $\left(\frac{2}{x} + \frac{1}{4x^2}\right)^{10}$.

Hence, or otherwise, obtain the term in $\frac{1}{x^{13}}$.   **5**

3. Use Gaussian elimination on the system of equations below to give an expression for z in terms of λ.

$$x + y + z = 2$$
$$4x + 3y - \lambda z = 4$$
$$5x + 6y + 8z = 11$$

For what values of λ does this system have a solution?

Determine the solution to this system of equations when $\lambda = 2$.   **6**

4. Given $x = \ln(1 + t^2)$, $y = \ln(1 + 2t^2)$ use parametric differentiation to find $\frac{dy}{dx}$ in terms of t.   **3**

5. Three vectors $\overrightarrow{OA}$, $\overrightarrow{OB}$ and $\overrightarrow{OC}$ are given by $\boldsymbol{u}$, $\boldsymbol{v}$ and $\boldsymbol{w}$ where

$$\boldsymbol{u} = 5\boldsymbol{i} + 13\boldsymbol{j}, \ \boldsymbol{v} = 2\boldsymbol{i} + \boldsymbol{j} + 3\boldsymbol{k}, \ \boldsymbol{w} = \boldsymbol{i} + 4\boldsymbol{j} - \boldsymbol{k}.$$

Calculate $\boldsymbol{u}.(\boldsymbol{v} \times \boldsymbol{w})$.   **3**

Interpret your result geometrically.   **1**

Marks

6. Given $e^y = x^3\cos^2 x$, $x > 0$, show that

$$\frac{dy}{dx} = \frac{a}{x} + b\tan x, \text{ for some constants } a \text{ and } b.$$

State the values of a and b. **3**

7. Given A is the matrix $\begin{pmatrix} 2 & a \\ 0 & 1 \end{pmatrix}$,

prove by induction that

$$A^n = \begin{pmatrix} 2^n & a\left(2^n - 1\right) \\ 0 & 1 \end{pmatrix}, \, n \geq 1.$$ **4**

8. Find the solution $y = f(x)$ to the differential equation

$$4\frac{d^2 y}{dx^2} - 4\frac{dy}{dx} + y = 0$$

given that $y = 4$ and $\dfrac{dy}{dx} = 3$ when $x = 0$. **6**

9. Give the first three non-zero terms of the Maclaurin series for $\cos 3x$. **2**

Write down the first four terms of the Maclaurin series for e^{2x}. **1**

Hence, or otherwise, determine the Maclaurin series for $e^{2x}\cos 3x$

up to, and including, the term in x^3. **3**

10. A semi-circle with centre $(1, 0)$ and radius 2, lies on the x-axis as shown.

Find the volume of the solid of revolution formed when the shaded region is rotated completely about the x-axis. **5**

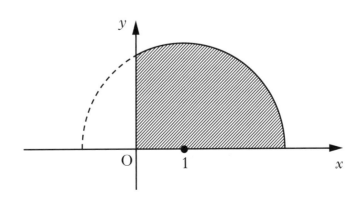

[Turn over

Mark

11. The function $f(x)$ is defined for all $x \geq 0$.

The graph of $y = f(x)$ intersects the y-axis at $(0, c)$, where $0 < c < 5$.

The graph of the function and its asymptote, $y = x - 5$, are shown below.

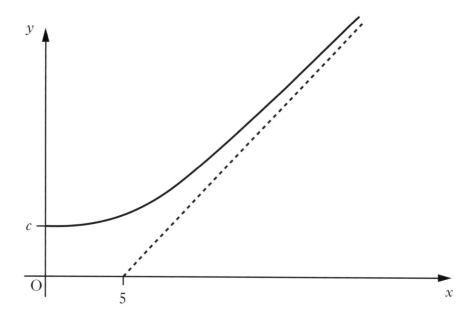

(a) Copy the above diagram.

On the same diagram, sketch the graph of $y = f^{-1}(x)$.

Clearly show any points of intersection and any asymptotes. **4**

(b) What is the equation of the asymptote of the graph of $y = f(x + 2)$? **1**

(c) Why does your diagram show that the equation $x = f(f(x))$ has at least one solution? **1**

12. Use the substitution $x = \tan\theta$ to determine the exact value of

$$\int_0^1 \frac{dx}{\left(1 + x^2\right)^{\frac{3}{2}}} .$$

6

13. The fuel efficiency, F, in km per litre, of a vehicle varies with its speed, s km per hour, and for a particular vehicle the relationship is thought to be

$$F = 15 + e^x(\sin x - \cos x - \sqrt{2}), \quad \text{where } x = \frac{\pi(s - 40)}{80},$$

for speeds in the range $40 \leq s \leq 120$ km per hour.

What is the greatest and least efficiency over the range and at what speeds do they occur? **10**

Marks

14. (a) Given the series $1 + r + r^2 + r^3 + \ldots$, write down the sum to infinity when $|r| < 1$.

Hence obtain an infinite geometric series for $\dfrac{1}{2 - 3r}$.

For what values of r is this series valid? **4**

(b) Express $\dfrac{1}{3r^2 - 5r + 2}$ in partial fractions.

Hence, or otherwise, determine the first three terms of an infinite series

for $\dfrac{1}{3r^2 - 5r + 2}$.

For what values of r does the series converge? **6**

15. (a) Use integration by parts to obtain an expression for

$$\int e^x \cos x \, dx.$$ **4**

(b) Similarly, given $I_n = \int e^x \cos nx \, dx$ where $n \neq 0$,

obtain an expression for I_n. **4**

(c) Hence evaluate $\int_0^{\frac{\pi}{2}} e^x \cos 8x \, dx$. **2**

16. (a) Express -1 as a complex number in polar form and hence determine the solutions to the equation $z^4 + 1 = 0$. **3**

(b) Write down the four solutions to the equation $z^4 - 1 = 0$. **2**

(c) Plot the solutions of both equations on an Argand diagram. **1**

(d) Show that the solutions of $z^4 + 1 = 0$ and the solutions of $z^4 - 1 = 0$ are also solutions of the equation $z^8 - 1 = 0$. **2**

(e) Hence identify all the solutions to the equation

$$z^6 + z^4 + z^2 + 1 = 0.$$ **2**

[*END OF QUESTION PAPER*]

[BLANK PAGE]

ADVANCED HIGHER | ANSWER SECTION

SQA ADVANCED HIGHER MATHEMATICS 2010–2014

ADVANCED HIGHER MATHEMATICS 2010

1. (a) For $f(x) = e^x \sin x^2$,
$$f'(x) = e^x \sin x^2 + e^x(2x \cos x^2).$$

(b) *Method 1*

For $g(x) = \dfrac{x^3}{(1 + \tan x)}$,

$$g'(x) = \frac{3x^2(1 + \tan x) - x^3 \sec^2 x}{(1 + \tan x)^2}.$$

Method 2

$$g(x) = x^3(1 + \tan x)^{-1}$$
$$g'(x) =$$
$$3x^2(1 + \tan x)^{-1} + x^3(-1)(1 + \tan x)^{-2} \sec^2 x$$
$$= \frac{x^2}{(1 + \tan x)^2}\left(3 + 3\tan x - x \sec^2 x\right)$$

2. Let the first term be a and the common ratio be r. Then
$$ar = -6 \qquad \text{and} \qquad ar^2 = 3$$
Hence
$$r = \frac{ar^2}{ar} = \frac{3}{-6} = -\frac{1}{2}.$$
So, since $|r| < 1$, the sum to infinity exists.
$$S = \frac{a}{1 - r}$$
$$= \frac{12}{1 - \left(-\frac{1}{2}\right)} = \frac{12}{\frac{3}{2}}$$
$$= 8.$$

3. (a)
$$t = x^4 \Rightarrow dt = 4x^3 dx$$
$$\int \frac{x^3}{1 + x^8} dx = \frac{1}{4}\int \frac{4x^3}{1 + (x^4)^2} dx$$
$$= \frac{1}{4}\int \frac{1}{1 + t^2} dt$$
$$= \tfrac{1}{4}\tan^{-1} t + c$$
$$= \tfrac{1}{4}\tan^{-1} x^4 + c$$

(b)
$$\int x^2 \ln x\, dx = \int(\ln x)\, x^2 dx$$
$$= \ln x \int x^2 dx - \int \frac{1}{x}\frac{x^3}{3} dx$$
$$= \tfrac{1}{3}x^3 \ln x - \tfrac{1}{3}\int x^2 dx$$
$$= \tfrac{1}{3}x^3 \ln x - \tfrac{1}{9}x^3 + c$$

4. The matrix $\begin{pmatrix} 2 & 0 \\ 0 & 2 \end{pmatrix}$ gives an enlargement, scale factor 2.

The matrix $\begin{pmatrix} \frac{1}{2} & \frac{\sqrt{3}}{2} \\ -\frac{\sqrt{3}}{2} & \frac{1}{2} \end{pmatrix}$ gives a clockwise rotation of 60° about the origin.

$$M = \begin{pmatrix} \frac{1}{2} & \frac{\sqrt{3}}{2} \\ -\frac{\sqrt{3}}{2} & \frac{1}{2} \end{pmatrix}\begin{pmatrix} 2 & 0 \\ 0 & 2 \end{pmatrix}$$
$$= \begin{pmatrix} 1 & \sqrt{3} \\ -\sqrt{3} & 1 \end{pmatrix}.$$

5.
$$\binom{n+1}{3} - \binom{n}{3} = \frac{(n+1)!}{3!\,(n-2)!} - \frac{n!}{3!\,(n-3)!}$$
$$= \frac{(n+1)!}{3!\,(n-2)!} - \frac{n!\,(n-2)}{3!\,(n-2)!}$$
$$= \frac{(n+1)! - n!\,(n-2)}{3!\,(n-2)!}$$
$$= \frac{n!\,[(n+1) - (n-2)]}{3!\,(n-2)!}$$
$$= \frac{n! \times 3}{3!\,(n-2)!} = \frac{n!}{2!\,(n-2)!}$$
$$= \binom{n}{2}$$

6.
$$\mathbf{v} \times \mathbf{w} = \begin{vmatrix} \mathbf{i} & \mathbf{j} & \mathbf{k} \\ 3 & 2 & -1 \\ -1 & 1 & 4 \end{vmatrix}$$
$$= \mathbf{i}\begin{vmatrix} 2 & -1 \\ 1 & 4 \end{vmatrix} - \mathbf{j}\begin{vmatrix} 3 & -1 \\ -1 & 4 \end{vmatrix} + \mathbf{k}\begin{vmatrix} 3 & 2 \\ -1 & 1 \end{vmatrix}$$
$$= 9\mathbf{i} - 11\mathbf{j} + 5\mathbf{k}$$
$$\mathbf{u} \cdot (\mathbf{v} \times \mathbf{w}) = (-2\mathbf{i} + 0\mathbf{j} + 5\mathbf{k}) \cdot (9\mathbf{i} - 11\mathbf{j} + 5\mathbf{k})$$
$$= -18 + 0 + 25$$
$$= 7.$$

7. $\int_1^2 \frac{3x+5}{(x+1)(x+2)(x+3)} dx$

$$\frac{3x+5}{(x+1)(x+2)(x+3)} = \frac{A}{x+1} + \frac{B}{x+2} + \frac{C}{x+3}$$
$$3x + 5 = A(x+2)(x+3) + B(x+1)(x+3) + C(x+1)(x+2)$$

$$x = -1 \Rightarrow 2 = 2A \Rightarrow A = 1$$
$$x = -2 \Rightarrow -1 = -B \Rightarrow B = 1$$
$$x = -3 \Rightarrow -4 = 2C \Rightarrow C = -2$$

Hence
$$\frac{3x+5}{(x+1)(x+2)(x+3)} = \frac{1}{x+1} + \frac{1}{x+2} - \frac{2}{x+3}$$
$$\int_1^2 \frac{3x+5}{(x+1)(x+2)(x+3)} dx = \int_1^2 \left(\frac{1}{x+1} + \frac{1}{x+2} - \frac{2}{x+3}\right) dx$$
$$= [\ln(x+1) + \ln(x+2) - 2\ln(x+3)]_1^2$$
$$= \ln 3 + \ln 4 - 2\ln 5 - \ln 2 - \ln 3 + 2\ln 4$$
$$= \ln\frac{3 \times 4 \times 4^2}{5^2 \times 2 \times 3} = \ln\frac{32}{25}$$

8. (a) Write the odd integers as: $2n + 1$ and $2m + 1$ where n and m are integers.
Then
$$(2n+1)(2m+1) = 4nm + 2n + 2m + 1$$
$$= 2(2nm + n + m) + 1$$
which is odd.

(b) Let $n = 1$, $p^1 = p$ which is given as odd. Assume p^k is odd and consider p^{k+1}.
$$p^{k+1} = p^k \times p$$
Since p^k is assumed to be odd and p is odd, p^{k+1} is the product of two odd integers is therefore odd.
Thus p^{n+1} is an odd integer for all n if p is an odd integer.

9. Let $f(x) = (1 + \sin^2 x)$. Then

$$f(0) = 1$$

$$f'(x) = 2\sin x \cos x \Rightarrow f'(0) = 0$$
$$= \sin 2x$$
$$f''(x) = 2\cos 2x \Rightarrow f''(0) = 2$$
$$f'''(x) = -4\sin 2x \Rightarrow f'''(0) = 0$$
$$f''''(x) = -8\cos 2x \Rightarrow f''''(0) = -8$$

$$f(x) = 1 + 2\frac{x^2}{2!} - 8\frac{x^4}{4!} + \ldots$$

$$= 1 + x^2 - \frac{1}{3}x^4 + \ldots$$

Alternative 1

$$f(0) = 1$$
$$f'(x) = 2\sin x \cos x \Rightarrow f'(0) = 0$$
$$f''(x) = 2\cos^2 x - 2\sin^2 x \Rightarrow f''(0) = 2$$
$$f'''(x) = 4(-\sin x)\cos x \Rightarrow f'''(0) = 0$$
$$-4\cos x \sin x$$
$$f''''(x) = -8\cos^2 x + 8\sin^2 x \Rightarrow f''''(0) = -8$$

etc

Alternative 2

$$f(x) = (1 + \sin^2 x)$$
$$= 1 + \tfrac{1}{2} - \tfrac{1}{2}\cos 2x$$
$$= \tfrac{1}{2}(3 - \cos 2x)$$
$$= \tfrac{1}{2}\left(3 - \left(1 - \frac{(2x)^2}{2!} + \frac{(2x)^4}{4!} - \ldots\right)\right)$$
$$= \tfrac{1}{2}\left(3 - 1 + 2x^2 - \tfrac{2}{3}x^4 - \ldots\right)$$
$$= 1 + x^2 - \tfrac{1}{3}x^4 - \ldots$$

10. The graph is not symmetrical about the y-axis (or $f(x) \neq f(-x)$) so it is not an even function.
The graph does not have half-turn rotational symmetry (or $f(x) \neq -f(-x)$) so it is not an odd function.
The function is neither even nor odd.

11.

$$\frac{d^2 y}{dx^2} + 4\frac{dy}{dx} + 5y = 0$$

$$m^2 + 4m + 5 = 0$$
$$(m + 2)^2 = -1$$
$$m = -2 \pm i$$

The general solution is

$$y = e^{-2x}(A\cos x + B\sin x)$$

$$x = 0, y = 3 \quad 3 = A$$
$$x = \tfrac{\pi}{2}, y = e^{-\pi} \Rightarrow e^{-\pi} = e^{-\pi}\left(3\cos\tfrac{\pi}{2} + B\sin\tfrac{\pi}{2}\right)$$
$$\Rightarrow B = 1$$

The particular solution is:

$$y = e^{-2x}(3\cos x + \sin x).$$

12. Assume $2 + x$ is rational

and let $2 + x = \dfrac{p}{q}$ where p, q are integers.

So

$$x = \frac{p}{q} - 2$$
$$= \frac{p - 2q}{q}$$

Since $p - 2q$ and q are integers, it follows that x is rational. This is a contradiction.

13.

$$y = t^3 - \tfrac{5}{2}t^2 \Rightarrow \frac{dy}{dt} = 3t^2 - 5t$$
$$x = \sqrt{t} = t^{1/2} \Rightarrow \frac{dx}{dt} = \tfrac{1}{2}t^{-1/2}$$

$$\Rightarrow \frac{dy}{dx} = \frac{3t^2 - 5t}{\tfrac{1}{2}t^{-1/2}}$$

$$= 6t^{5/2} - 10t^{3/2}$$

$$\frac{d^2 y}{dx^2} = \frac{\frac{d}{dt}\left(\frac{dy}{dx}\right)}{\frac{dx}{dt}}$$

$$= \frac{6 \times \tfrac{5}{2}t^{3/2} - 10 \times \tfrac{3}{2}t^{1/2}}{\tfrac{1}{2}t^{-1/2}}$$

$$= 30t^2 - 30t$$

i.e. $a = 30, \ b = -30$

At a point of inflexion, $\dfrac{d^2 y}{dx^2} = 0 \Rightarrow t = 0$ or 1
But $t > 0 \Rightarrow t = 1 \Rightarrow \dfrac{dy}{dx} = -4$
and the point of contact is $\left(1, -\tfrac{3}{2}\right)$
Hence the tangent is

$$y + \tfrac{3}{2} = -4(x - 1)$$
i.e. $2y + 8x = 5$

14.

$$\begin{array}{ccc|c} 1 & -1 & 1 & 1 \\ 1 & 1 & 2 & 0 \\ 2 & -1 & a & 2 \end{array}$$

$$\begin{array}{ccc|c} 1 & -1 & 1 & 1 \\ 0 & 2 & 1 & -1 \\ 0 & 1 & a-2 & 0 \end{array}$$

$$\begin{array}{ccc|c} 1 & -1 & 1 & 1 \\ 0 & 2 & 1 & -1 \\ 0 & 0 & 2a-5 & 1 \end{array}$$

$$z = \frac{1}{2a - 5};$$

$$2y + \frac{1}{2a - 5} = -1 \Rightarrow 2y = \frac{-2a + 5 - 1}{2a - 5}$$

$$\Rightarrow y = \frac{2 - a}{2a - 5};$$

$$x - \frac{2 - a}{2a - 5} + \frac{1}{2a - 5} = 1$$

$$\Rightarrow x = \frac{2a - 5}{2a - 5} + \frac{1 - a}{2a - 5} = \frac{a - 4}{2a - 5}.$$

which exist when $2a - 5 \neq 0$.

From the third row of the final tableau, when $a = 2.5$, there are no solutions

When $a = 3, x = -1, y = -1, z = 1$.

$$AB = \begin{pmatrix} 5 & 2 & -3 \\ 1 & 1 & -1 \\ -3 & -1 & 2 \end{pmatrix}\begin{pmatrix} 1 \\ 0 \\ 2 \end{pmatrix} = \begin{pmatrix} -1 \\ -1 \\ 1 \end{pmatrix}$$

From above, we have $C\begin{pmatrix} -1 \\ -1 \\ 1 \end{pmatrix} = \begin{pmatrix} 1 \\ 0 \\ 2 \end{pmatrix}$ and

also $A\begin{pmatrix} 1 \\ 0 \\ 2 \end{pmatrix} = \begin{pmatrix} -1 \\ -1 \\ 1 \end{pmatrix}$ which suggests $AC = I$ and

this can be verified directly. Hence
A is the inverse of C (or vice versa).

15. $(x^2)^2 = 8x \Rightarrow x^4 = 8x \Rightarrow x = 0, 2$

$$\text{Area} = 4\int_0^2 (\sqrt{8x} - x^2)\, dx$$

$$= 4\left[\sqrt{8}\left(\frac{2}{3}x^{3/2}\right) - \frac{1}{3}x^3\right]_0^2$$

$$= 4\left[\frac{16}{3} - \frac{8}{3}\right] = \frac{32}{3}$$

Volume of revolution about the y-axis $= \pi\int x^2 dy$.
So in this case, we need to calculate
two volumes and subtract:

$$V = \pi\left[\int_0^4 y\, dy\right] - \pi\left[\int_0^4 \frac{y^4}{64}dy\right]$$

$$= \pi\left[\frac{y^2}{2}\right]_0^4 - \pi\left[\frac{y^5}{320}\right]_0^4$$

$$= \pi\left[8 - \frac{64 \times 4^2}{320}\right]$$

$$= \frac{40 - 16}{5}\pi$$

$$= \frac{24\pi}{5} \quad (15)$$

16. $z^3 = r^3(\cos 3\theta + i \sin 3\theta)$

$\left(\cos\frac{2\pi}{3} + i \sin\frac{2\pi}{3}\right)^3 = \cos 2\pi + i \sin 2\pi$

$a = 1; \ b = 0$

Method 1

$r^3(\cos 3\theta + i \sin 3\theta) = 8$

$r^3 \cos 3\theta = 8$ and $r^3 \sin 3\theta = 0$

$\Rightarrow r = 2; \ 3\theta = 0, 2\pi, 4\pi$

Roots are $2, 2\left(\cos\frac{2\pi}{3} + i\sin\frac{2\pi}{3}\right), 2\left(\cos\frac{4\pi}{3} + i\sin\frac{4\pi}{3}\right)$.
In cartesian form: $2, (-1 + i\sqrt{3}), (-1 - i\sqrt{3})$

Method 2

$$z^3 - 8 = 0$$
$$(z - 2)(z^2 + 2z + 4) = 0$$
$$(z - 2)((z + 1)^2 + (\sqrt{3})^2) = 0$$

so the roots are: $2, (-1 + i\sqrt{3}), (-1 - i\sqrt{3})$

(a) $z_1 + z_2 + z_3 = 0$

(b) Since $z_1^3 = z_2^3 = z_3^3 = 8$

it follows that

$$z_1^6 + z_2^6 + z_3^6 = (z_1^3)^2 + (z_2^3)^2 + (z_3^3)^2$$
$$= 3 \times 64 = 192$$

---

1. $$\frac{13 - x}{x^2 + 4x - 5} = \frac{13 - x}{(x - 1)(x + 5)}$$

$$= \frac{A}{x - 1} + \frac{B}{x + 5}$$

$13 - x = A(x + 5) + B(x - 1)$

$x = 1 \Rightarrow 12 = 6A \Rightarrow A = 2$

$x = -5 \Rightarrow 18 = -6B \Rightarrow B = -3$

Hence $\dfrac{13 - x}{x^2 + 4x - 5} = \dfrac{2}{x - 1} - \dfrac{3}{x + 5}$

$$\int \frac{13 - x}{x^2 + 4x - 5}dx = \int \frac{2}{x - 1}dx - \int \frac{3}{x + 5}dx$$

$$= 2 \ln|x - 1| - 3 \ln|x + 5| + c$$

2. $\left(\frac{1}{2}x - 3\right)^4 = {}^4C_0\left(\frac{x}{2}\right)^4 + {}^4C_1\left(\frac{x}{2}\right)^3(-3) + $
$\left. {}^4C_2\left(\frac{x}{2}\right)^2(-3)^2 + {}^4C_3\left(\frac{x}{2}\right)(-3)^3 + {}^4C_4(-3)^4\right\}$

$= \left(\frac{x}{2}\right)^4 + 4\left(\frac{x}{2}\right)^3(-3) + 6\left(\frac{x}{2}\right)^2(-3)^2 + 4\left(\frac{x}{2}\right)(-3)^3 + (-3)^4$

$= \dfrac{x^4}{16} - \dfrac{3x^3}{2} + \dfrac{27x^2}{2} - 54x + 81.$

3.   (a) *Method 1*

$$y + e^y = x^2$$

$$\frac{dy}{dx} + e^y\frac{dy}{dx} = 2x$$

$$\frac{dy}{dx}(1 + e^y) = 2x \Rightarrow \frac{dy}{dx} = \frac{2x}{(1 + e^y)}$$

Method 2

$$\ln(y + e^y) = 2\ln x$$
$$\frac{(1 + e^y)\frac{dy}{dx}}{y + e^y} = \frac{2}{x}$$
$$\frac{dy}{dx} = \frac{2(y + e^y)}{x(1 + e^y)}$$

Method 3

$y + e^y = x^2 \Rightarrow e^y = x^2 - y \Rightarrow y = \ln(x^2 - y)$

$$\frac{dy}{dx} = \frac{2x - \frac{dy}{dx}}{x^2 - y}$$

$$\frac{dy}{dx}(x^2 - y) = 2x - \frac{dy}{dx} \Rightarrow \frac{dy}{dx}(x^2 - y + 1) = 2x$$

$$\frac{dy}{dx} = \frac{2x}{x^2 - y + 1}$$

(b) *Method 1*

$f(x) = \sin x \cos^3 x$

$f'(x) = \cos^4 x + \sin x(-3\cos^2 x \sin x)$

$= \cos^4 x - 3\cos^2 x \sin^2 x$

Method 2

$f(x) = \sin x \cos^3 x$

$\ln(f(x)) = \ln \sin x + \ln(\cos^3 x)$

$\dfrac{f'(x)}{f(x)} = \dfrac{\cos x}{\sin x} - \dfrac{3\cos^2 x \sin x}{\cos^3 x}$

$= \dfrac{\cos x}{\sin x} - \dfrac{3\sin x}{\cos x}$

$f'(x) = \left(\dfrac{\cos x}{\sin x} - \dfrac{3\sin x}{\cos x}\right)\sin x \cos^3 x$

$= \cos^4 x - 3\sin^2 x \cos^2 x$

4. (*a*)   Singular when the determinant is 0.

$$1\det\begin{pmatrix}0 & 2\\ \lambda & 6\end{pmatrix} - 2\det\begin{pmatrix}3 & 2\\ -1 & 6\end{pmatrix} + (-1)\det\begin{pmatrix}3 & 0\\ -1 & \lambda\end{pmatrix} = 0$$

$$-2\lambda - 2(18 + 2) - 1(3\lambda - 0) = 0$$

$$-5\lambda - 40 = 0 \text{ when } \lambda = -8$$

(*b*)   From A, $A' = \begin{vmatrix}2 & 3\alpha + 2\beta & -1\\ 2\alpha - \beta & 4 & 3\\ -1 & 3 & 2\end{vmatrix}$.

Hence $2\alpha - \beta = -1$ and $3\alpha + 2\beta = -5$.

$$4\alpha - 2\beta = -2$$

$$3\alpha + 2\beta = -5$$

$$7\alpha = -7 \Rightarrow \alpha = -1 \text{ and } \beta = -1.$$

5.   Let $f(x) = (1 + x)^{\frac{1}{2}}$, then

$$f(x) = (1 + x)^{\frac{1}{2}} \Rightarrow f(0) = 1$$

$$f'(x) = \tfrac{1}{2}(1 + x)^{-\frac{1}{2}} \Rightarrow f'(0) = \tfrac{1}{2}$$

$$f''(x) = -\tfrac{1}{4}(1 + x)^{-\frac{3}{2}} \Rightarrow f''(0) = -\tfrac{1}{4}$$

$$f'''(x) = \tfrac{3}{8}(1 + x)^{-\frac{5}{2}} \Rightarrow f'''(0) = \tfrac{3}{8}$$

Hence

$$\sqrt{1 + x} = 1 + \tfrac{1}{2}x - \tfrac{1}{4} \times \tfrac{x^2}{2} + \tfrac{3}{8} \times \tfrac{x^3}{6} - \ldots$$

$$= 1 + \tfrac{1}{2}x - \tfrac{x^2}{8} + \tfrac{x^3}{16} - \ldots$$

and replacing x by x^2 gives

$$\sqrt{1 + x^2} = 1 + \tfrac{1}{2}x^2 - \tfrac{x^4}{8} + \tfrac{x^6}{16} - \ldots$$

Thus

$$\sqrt{(1 + x)(1 + x^2)} =$$

$$\left(1 + \tfrac{1}{2}x - \tfrac{x^2}{8} + \tfrac{x^3}{16} - \ldots\right)\left(1 + \tfrac{1}{2}x^2 - \tfrac{x^4}{8} + \tfrac{x^6}{16} - \ldots\right)$$

$$= 1 + \tfrac{1}{2}x + \tfrac{1}{2}x^2 - \tfrac{1}{8}x^2 + \tfrac{1}{4}x^3 + \tfrac{1}{16}x^3 + \ldots$$

$$= 1 + \tfrac{1}{2}x + \tfrac{3}{8}x^2 + \tfrac{5}{16}x^3 + \ldots$$

6. Reflect in the line $y = x$ to get

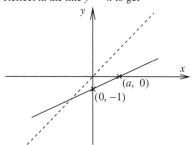

Now apply the modulus function

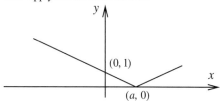

7. *Method 1*

$$y = \frac{e^{\sin x}(2 + x)^3}{\sqrt{1 - x}}$$

$$\Rightarrow \ln y = \ln\left(e^{\sin x}(2 + x)^3\right) - \ln\left(\sqrt{1 - x}\right)$$

$$= \sin x + 3\ln(2 + x) - \tfrac{1}{2}\ln(1 - x)$$

$$\Rightarrow \frac{1}{y}\frac{dy}{dx} = \cos x + \frac{3}{2 + x} + \frac{1}{2(1 - x)}$$

$$\frac{dy}{dx} = y\left(\cos x + \frac{3}{2 + x} + \frac{1}{2(1 - x)}\right)$$

When $x = 0$, $y = 8 \Rightarrow$

gradient $= 8\left(1 + \dfrac{3}{2} + \dfrac{1}{2}\right) = 24.$

Method 2

$$y = \frac{e^{\sin x}(2 + x)^3}{\sqrt{1 - x}} \Rightarrow$$

$$\frac{dy}{dx} = \frac{\frac{d}{dx}\left(e^{\sin x}(2+x)^3\right)\sqrt{1-x} - e^{\sin x}(2+x)^3\left(-\frac{1}{2}\frac{1}{\sqrt{1-x}}\right)}{(1 - x)}$$

$$= \frac{\left[\cos x\, e^{\sin x}(2+x)^3 + 3e^{\sin x}(2+x)^2\right](1-x)}{(1 - x)^{3/2}} + \frac{e^{\sin x}(2 + x)^3}{2(1 - x)^{3/2}}$$

When $x = 0$, gradient $= \dfrac{(2^3 + 3 \times 2^2)}{1} + \dfrac{2^3}{2} = 20 + 4 = 24$

Method 3

$$y = \frac{e^{\sin x}(2 + x)^3}{\sqrt{1 - x}}$$

$$y\sqrt{1 - x} = e^{\sin x}(2 + x)^3$$

$$\sqrt{1 - x}\tfrac{dy}{dx} - \tfrac{1}{2}y(1 - x)^{-1/2} = \cos x\, e^{\sin x}(2 + x)^3 + 3e^{\sin x}(2 + x)^2$$

when $x = 0$, $y = \dfrac{e^0 2^3}{1} = 8$. This leads to $\tfrac{dy}{dx} = 24$

8.

$$\sum_{r=1}^{n} r^3 - \left(\sum_{r=1}^{n} r\right)^2 = \frac{n^2(n + 1)^2}{4} - \left(\frac{n(n + 1)}{2}\right)^2 = 0$$

$$\sum_{r=1}^{n} r^3 + \left(\sum_{r=1}^{n} r\right)^2 = \frac{n^2(n + 1)^2}{4} + \left(\frac{n(n + 1)}{2}\right)^2$$

$$= \frac{n^2(n + 1)^2}{4} + \frac{n^2(n + 1)^2}{4}$$

$$= \frac{n^2(n + 1)^2}{2}$$

9. *Method 1*

$$\frac{dy}{dx} = 3(1 + y)\sqrt{1 + x}$$

$$\int\frac{dy}{1 + y} = 3\int(1 + x)^{\frac{1}{2}}\,dx$$

$$\ln(1 + y) = 2(1 + x)^{\frac{3}{2}} + c$$

$$1 + y = \exp\left(2(1 + x)^{\frac{3}{2}} + c\right)$$

$$y = \exp\left(2(1 + x)^{\frac{3}{2}} + c\right) - 1.$$

$$= A\exp\left(2(1 + x)^{\frac{3}{2}}\right) - 1.$$

Method 2

$$\frac{dy}{dx} - 3\sqrt{1 + x}\,y = 3\sqrt{1 + x}$$

Integrating Factor

$$\exp\left(-3\int\sqrt{1 + x}\,dx\right) = \exp\left(-2(1 + x)^{3/2}\right)$$

$$\tfrac{d}{dx}\left(y\exp\left(-2(1 + x)^{3/2}\right)\right) = 3\sqrt{1 + x}\left(\exp\left(-2(1 + x)^{3/2}\right)\right)$$

$$y\left(\exp\left(-2(1 + x)^{3/2}\right)\right) = -\int -3\sqrt{1 + x}\exp\left(-2(1 + x)^{3/2}\right)dx$$

$$= -\exp\left(-2(1 + x)^{3/2}\right) + c$$

$$y = -1 + c\exp\left(2(1 + x)^{3/2}\right)$$

10. Let $z = x + iy$ so $z - 1 = (x - 1) + iy$.
$|z - 1|^2 = (x - 1)^2 + y^2 = 9$.
The locus is the circle with centre $(1, \ 0)$ and radius 3.

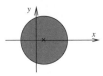

11. (a) $\int_0^{\pi/4}(\sec x - x)(\sec x + x)dx = \int_0^{\pi/4}(\sec^2 x - x^2)dx$

$$= \left[\tan x - \frac{x^3}{3}\right]_0^{\frac{\pi}{4}}$$

$$= \left[1 - \frac{1}{3}\frac{\pi^3}{64}\right] - [0]$$

$$= 1 - \frac{\pi^3}{192}.$$

(b) *Method 1*

Let $u = 7x^2$, then $du = 14x\, dx$.

$$\int \frac{x}{\sqrt{1 - 49x^4}}\, dx = \frac{1}{14}\int \frac{du}{\sqrt{1 - u^2}}$$

$$= \frac{1}{14}\sin^{-1}u + c$$

$$= \frac{1}{14}\sin^{-1}7x^2 + c$$

Method 2

$$\int \frac{x}{\sqrt{1 - 49x^4}}\, dx = \frac{1}{14}\int \frac{14x\, dx}{\sqrt{1 - (7x^2)^2}}$$

$$= \frac{1}{14}\sin^{-1}7x^2 + c$$

12. For $n = 2$, $8^2 + 3^0 = 64 + 1 = 65$.
True when $n = 2$.

Assume true for k, i.e. that $8^k + 3^{k-2}$ is divisible by 5, i.e. can be expressed as $5p$ for an integer p.

Now consider $8^{k+1} + 3^{k-1}$
$= 8 \times 8^k + 3^{k-1}$
$= 8 \times (5p - 3^{k-2}) + 3^{k-1}$
$= 40p - 3^{k-2}(8 - 3)$
$= 5(8p - 3^{k-2})$ which is divisible by 5.
So, since it is true for $n = 2$, it is true for all $n \geqslant 2$.

13. *Method 1*

Let d be the common difference. Then
$u_3 = 1 = a + 2d$ and $u_2 = \frac{1}{a} = a + d$

$$1 = a + 2\left(\frac{1}{a} - a\right)$$

$$a = a^2 + 2 - 2a^2$$

$a^2 + a - 2 = 0$
$(a + 2)(a - 1) = 0 \Rightarrow a = -2$ since $a < 0$.
$a = -2$ gives $2d = 3$ and hence $d = \frac{3}{2}$.

Method 2

$u_1 = a, u_2 = \frac{1}{a}, u_3 = 1$

$$\Rightarrow \frac{1}{a} - a = 1 - \frac{1}{a}$$

$$\Rightarrow 1 - a^2 = a - 1$$

$$\Rightarrow a^2 + a - 2 = 0$$

$(a + 2)(a - 1) = 0 \Rightarrow a = -2$ since $a < 0$.

$$d = u_3 - u_2 = 1 - \frac{1}{a} = \frac{3}{2}$$

13. (*continued*)

$$S_n = \frac{n}{2}[2a + (n - 1)d]$$

$$= \frac{n}{2}\left[-4 + \frac{3}{2}n - \frac{3}{2}\right]$$

$$= \tfrac{1}{4}[3n^2 - 11n]$$

$$\therefore 3n^2 - 11n > 4000$$

$$n^2 - \frac{11}{3}n > \frac{4000}{3}$$

$$\left(n - \frac{11}{6}\right)^2 > \frac{48000}{36} + \frac{121}{36} = \frac{48121}{36}$$

$$n - \frac{11}{6} > \frac{\sqrt{48121}}{6}$$

$$n > \frac{\sqrt{48121} + 11}{6} \approx 38.39$$

So the least value of n is 39.

14. Auxiliary equation
$$m^2 - m - 2 = 0$$
$$(m - 2)(m + 1) = 0$$
$$m = -1 \text{ or } 2$$
Complementary function is: $y = Ae^{-x} + Be^{2x}$

The particular integral has the form $y = Ce^x + D$
$$y = Ce^x + D \Rightarrow \frac{dy}{dx} = Ce^x$$
$$\Rightarrow \frac{d^2y}{dx^2} = Ce^x$$
Hence we need:
$$\frac{d^2y}{dx^2} - \frac{dy}{dx} - 2y = e^x + 12$$
$$[Ce^x] - [Ce^x] - 2[Ce^x + D] = e^x + 12$$
$$-2Ce^x - 2D = e^x + 12$$
Hence $C = -\frac{1}{2}$ and $D = -6$.
So the General Solution is
$$y = Ae^{-x} + Be^{2x} - \tfrac{1}{2}e^x - 6.$$
$x = 0$ and $y = -\frac{3}{2} \Rightarrow A + B - \frac{1}{2} - 6 = -\frac{3}{2}$
$x = 0$ and $\frac{dy}{dx} = \frac{1}{2} \Rightarrow -A + 2B - \frac{1}{2} = \frac{1}{2}$
$$3B - 7 = -1 \Rightarrow B = 2 \Rightarrow A = 3$$
So the particular solution is
$$y = 3e^{-x} + 2e^{2x} - \tfrac{1}{2}e^x - 6.$$

15. (a) In terms of a parameter s, L_1 is given by
$$x = 1 + ks, \ y = -s, \ z = -3 + s$$

In terms of a parameter t, L_2 is given by
$$x = 4 + t, \ y = -3 + t, \ z = -3 + 2t$$

Equating the y coordinates and equating the z coordinates:

$$-s = -3 + t$$
$$-3 + s = -3 + 2t$$
Adding these
$$-3 = -6 + 3t \Rightarrow t = 1 \Rightarrow s = 2.$$

From the x coordinates
$$1 + ks = 4 + t$$
Using the values of s and t
$$1 + 2k = 5 \Rightarrow k = 2$$

The point of intersection is: $(5, -2, -1)$.

(b) L_1 has direction $2\mathbf{i} - \mathbf{j} + \mathbf{k}$.
L_2 has direction $\mathbf{i} + \mathbf{j} + 2\mathbf{k}$.

Let the angle between L_1 and L_2 be θ, then
$$\cos\theta = \frac{(2\mathbf{i} - \mathbf{j} + \mathbf{k}).(\mathbf{i} + \mathbf{j} + 2\mathbf{k})}{|2\mathbf{i} - \mathbf{j} + \mathbf{k}||\mathbf{i} + \mathbf{j} + 2\mathbf{k}|}$$
$$= \frac{2 - 1 + 2}{\sqrt{6}\sqrt{6}} = \frac{3}{6} = \frac{1}{2}$$
$$\theta = 60°$$
The angle between L_1 and L_2 is $60°$.

16. (a) $I_n = \int_0^1 \frac{1}{(1 + x^2)^n}\, dx$

$= \int_0^1 1 \times (1 + x^2)^{-n}\, dx$

$= \left[(1 + x^2)^{-n}\int 1\, dx\right]_0^1 + \int_0^1 \left(2nx(1 + x^2)^{-n-1}\int 1\, dx\right) dx$

$= \left[x(1 + x^2)^{-n}\right]_0^1 + \int_0^1 2nx^2(1 + x^2)^{-n-1}\, dx$

$= \frac{1}{2^n} - 0 + 2n\int_0^1 x^2 (1 + x^2)^{-n-1}\, dx$

$= \frac{1}{2^n} + 2n\int_0^1 \frac{x^2}{(1 + x^2)^{n+1}}\, dx.$

(b) $\dfrac{A}{(1 + x^2)^n} + \dfrac{B}{(1 + x^2)^{n+1}} = \dfrac{x^2}{(1 + x^2)^{n+1}}$

$\Rightarrow A(1 + x^2) + B = x^2$

$\Rightarrow A = 1,\ B = -1$

$\dfrac{1}{(1 + x^2)^n} + \dfrac{-1}{(1 + x^2)^{n+1}} = \dfrac{x^2}{(1 + x^2)^{n+1}}$

$I_n = \dfrac{1}{2^n} + 2n\int_0^1 \dfrac{x^2}{(1 + x^2)^{n+1}}\, dx.$

$= \dfrac{1}{2^n} + 2n\int_0^1 \dfrac{1}{(1 + x^2)^n}\, dx + 2n\int_0^1 \dfrac{-1}{(1 + x^2)^{n+1}}\, dx$

$= \dfrac{1}{2^n} + 2nI_n - 2nI_{n+1}$

$2nI_{n+1} = \dfrac{1}{2^n} + (2n - 1)I_n$

$I_{n+1} = \dfrac{1}{n \times 2^{n+1}} + \left(\dfrac{2n - 1}{2n}\right)I_n.$

(c) $\int_0^1 \dfrac{1}{(1 + x^2)^3}\, dx = I_3$

$= \dfrac{1}{16} + \dfrac{3}{4}I_2$

$= \dfrac{1}{16} + \dfrac{3}{4}\left(\dfrac{1}{4} + \dfrac{1}{2}I_1\right)$

$= \dfrac{1}{4} + \dfrac{3}{8}\int_0^1 \dfrac{1}{1 + x^2}\, dx$

$= \dfrac{1}{4} + \dfrac{3}{8}\left[\tan^{-1}x\right]_0^1$

$= \dfrac{1}{4} + \dfrac{3}{8}\dfrac{\pi}{4} = \dfrac{1}{4} + \dfrac{3\pi}{32}.$

ADVANCED HIGHER MATHEMATICS 2012

1. (a) $f(x) = \dfrac{3x + 1}{x^2 + 1}$

$f'(x) = \dfrac{3(x^2 + 1) - (3x + 1)2x}{(x^2 + 1)^2}$

$= \dfrac{3x^2 + 3 - 6x^2 - 2x}{(x^2 + 1)^2}$

$= \dfrac{-3x^2 - 2x + 3}{(x^2 + 1)^2}$

(b) $g(x) = \cos^2 x\, e^{\tan x}$

$g'(x) = 2\cos x(-\sin x)e^{\tan x} + (\cos^2 x)(\sec^2 x)e^{\tan x}$

$= -\sin 2x\, e^{\tan x} + e^{\tan x}$

$= (1 - \sin 2x)e^{\tan x}$

2. $a = 2048$ and $ar^3 = 256$

$\Rightarrow r^3 = \dfrac{1}{8}$

$\Rightarrow r = \dfrac{1}{2}.$

$S_n = \dfrac{a(1 - r^n)}{1 - r}$

$\Rightarrow \dfrac{1 - (\frac{1}{2})^n}{1 - \frac{1}{2}} = \dfrac{4088}{2048}$

$= \dfrac{511}{256}$

$\Rightarrow 1 - \left(\dfrac{1}{2}\right)^n = \dfrac{511}{256} \times \dfrac{1}{2} = \dfrac{511}{512}$

$\dfrac{1}{2^n} = 1 - \dfrac{511}{512} = \dfrac{1}{512}$

$\Rightarrow 2^n = 512 \Rightarrow n = 9$

3. Since w is a root, $\overline{w} = -1 - 2i$ is also a root.
The corresponding factors are
$(z + 1 - 2i)$ and $(z + 1 + 2i)$
from which
$((z + 1) - 2i)((z + 1) + 2i) = (z + 1)^2 + 4$
$= z^2 + 2z + 5$
$z^3 + 5z^2 + 11z + 15 = (z^2 + 2z + 5)(z + 3)$

The roots are $(-1 + 2i)$, $(-1 - 2i)$ and -3.

4. The general term is given by:
$$\binom{9}{r}(2x)^{9-r}\left(-\frac{1}{x^2}\right)^r$$

$= \binom{9}{r} \times \dfrac{2^{9-r}x^{9-r}(-1)^r}{x^{2r}}$

$= \binom{9}{r} \times (-1)^r 2^{9-r}x^{9-3r}$

The term independent of x occurs when
$9 - 3r = 0$, i.e. when $r = 3$.

The term is: $\dfrac{9!}{6!\,3!}(-1)^3 2^6$

$= -5376.$

5. *Method 1*

$$\overrightarrow{PQ} = 3\mathbf{i} + \mathbf{j} + 4\mathbf{k} \text{ and } \overrightarrow{QR} = 2\mathbf{i} - 2\mathbf{j} - 2\mathbf{k}$$

A normal to the plane:

$$\overrightarrow{PQ} \times \overrightarrow{QR} = \begin{vmatrix} \mathbf{i} & \mathbf{j} & \mathbf{k} \\ 3 & 1 & 4 \\ 2 & -2 & -2 \end{vmatrix}$$

$$= \mathbf{i}\begin{vmatrix} 1 & 4 \\ -2 & -2 \end{vmatrix} - \mathbf{j}\begin{vmatrix} 3 & 4 \\ 2 & -2 \end{vmatrix} + \mathbf{k}\begin{vmatrix} 3 & 1 \\ 2 & -2 \end{vmatrix}$$

$$= 6\mathbf{i} + 14\mathbf{j} - 8\mathbf{k}$$

Hence the equation has the form:

$$6x + 14y - 8z = d.$$

The plane passes through $P(-2, 1, -1)$ so

$$d = -12 + 14 + 8 = 10$$

which gives an equation $6x + 14y - 8z = 10$

i.e. $3x + 7y - 4z = 5$.

Method 2

A plane has an equation of the form $ax + by + cz = d$.

Using the points P, Q, R we get

$$-2a + b - c = d$$
$$a + 2b + 3c = d$$
$$3a + c = d$$

Using Gaussian elimination to solve these we have

$$\begin{vmatrix} -2 & 1 & -1 & d \\ 1 & 2 & 3 & d \\ 3 & 0 & 1 & d \end{vmatrix} \Rightarrow \begin{vmatrix} -2 & 1 & -1 & d \\ 0 & 5 & 5 & 3d \\ 0 & 6 & 8 & 2d \end{vmatrix}$$

$$\Rightarrow \begin{vmatrix} -2 & 1 & -1 & d \\ 0 & 5 & 5 & 3d \\ 0 & 0 & 2 & -\frac{8}{5}d \end{vmatrix}$$

$$\Rightarrow c = -\frac{4}{5}d, \qquad b = \frac{7}{5}d, \qquad a = \frac{3}{5}d$$

These give the equation

$$\left(\tfrac{3}{5}d\right)x + \left(\tfrac{7}{5}d\right)y + \left(-\tfrac{4}{5}d\right)z = d$$

i.e. $\quad 3x + 7y - 4z = 5$

6. *Method 1*

A $\quad e^x = 1 + x + \frac{x^2}{2} + \frac{x^3}{6} + \dots$

B $\quad (1 + e^x)^2 = 1 + 2e^x + e^{2x}$

C $\quad = 1 + 2\left(1 + x + \frac{x^2}{2} + \frac{x^3}{6} + \dots\right) + \left(1 + 2x + \frac{(2x)^2}{2} + \frac{(2x)^3}{6} + \dots\right)$

D $\quad = 1 + 2 + 2x + x^2 + \frac{1}{3}x^3 + 1 + 2x + 2x^2 + \frac{4}{3}x^3 + \dots$

E $\quad = 4 + 4x + 3x^2 + \frac{5}{3}x^3 + \dots$

Method 2

$$e^x = 1 + x + \frac{x^2}{2} + \frac{x^3}{6} + \dots$$

$$(1 + e^x) = 2 + x + \frac{x^2}{2} + \frac{x^3}{6} + \dots$$

$$(1 + e^x)^2 = \left(2 + x + \frac{x^2}{2} + \frac{x^3}{6} + \dots\right)\left(2 + x + \frac{x^2}{2} + \frac{x^3}{6} + \dots\right)$$

$$= 4 + 4x + 3x^2 + \frac{1}{3}x^3 + \frac{1}{2}x^3 + \frac{1}{2}x^3 + \frac{1}{3}x^3 + \dots$$

$$= 4 + 4x + 3x^2 + \frac{5}{3}x^3 + \dots$$

Method 3

$$e^x = 1 + x + \frac{x^2}{2} + \frac{x^3}{6} + \dots$$

$$f(x) = (1 + e^x)^2 \qquad f(0) = 4$$

$$f'(x) = 2e^x(1 + e^x) \qquad f'(0) = 4$$

$$= 2e^x + 2e^{2x}$$

$$f''(x) = 2e^x + 4e^{2x} \qquad f''(0) = 6$$

$$f'''(x) = 2e^x + 8e^{2x} \qquad f'''(0) = 10$$

$$f(x) = f(0) + f'(0)x + f''(0)\frac{x^2}{2} + f'''(0)\frac{x^3}{6} + \dots$$

$$(1 + e^x)^2 = 4 + 4x + 3x^2 + \frac{5}{3}x^3 + \dots$$

7. *(a)*

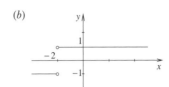

$$y = |x + 2|$$

(b)

8. $\quad x = 4\sin\theta \Rightarrow dx = 4\cos\theta\,d$

$$\int_0^2 \sqrt{16 - x^2}\,dx$$

$$= \int_0^{\pi/6} \sqrt{16 - (4\sin\theta)^2}\,.\,4\cos\theta\,d\theta$$

$$= \int_0^{\pi/6} \sqrt{16(1 - \sin^2\theta)}\,.\,4\cos\theta\,d\theta$$

$$= \int_0^{\pi/6} \sqrt{16\cos^2\theta}\,.\,4\cos\theta\,d$$

$$= \int_0^{\pi/6} 16\cos^2\theta\,d\theta$$

$$= 8\int_0^{\pi/6}(1 + \cos 2\theta)\,d\theta$$

$$= 8\left[\theta + \tfrac{1}{2}\sin 2\theta\right]_0^{\pi/6}$$

$$= \tfrac{8\pi}{6} + 4\sin\tfrac{\pi}{3}$$

$$= \tfrac{4\pi}{3} + 2\sqrt{3}\,(\approx 7.65)$$

9. *Method 1*

$$A + A^{-1} = I$$

$$A^2 + I = A$$

Hence $\quad A^2 + I = I - A^{-1}$

$$A^2 = -A^{-1}$$

$$A^3 = -I, \text{ i.e. } k = -1$$

Method 2

$$A + A^{-1} = I$$

$$A = I - A^{-1}$$

$$A^2 = I - 2A^{-1} + (A^{-1})^2$$

$$A^3 = A - 2I + A^{-1}$$

$$A^3 = (A + A^{-1}) - 2I = I - 2I$$

Hence $\quad A^3 = -I, \text{ i.e. } k = -1$

10. *Method 1*

$$1234 = 7 \times 176 + 2$$
$$176 = 7 \times 25 + 1$$
$$25 = 7 \times 3 + 4$$

Hence

$$1234_{10} = 3412_7$$

Method 2

$$1234 = 7 \times 176 + 2$$
$$= 7 \times (7 \times 25 + 1) + 2$$
$$= 7 \times (7 \times (7 \times 3 + 4) + 1) + 2$$
$$= 3 \times 7^3 + 4 \times 7^2 + 1 \times 7 + 2$$

Hence

$$1234_{10} = 3412_7$$

11. (a) $\dfrac{d}{dx}\left(\sin^{-1}x\right) = \dfrac{1}{\sqrt{1-x^2}}$

(b) $\int \sin^{-1}x . \dfrac{x}{\sqrt{1-x^2}}\,dx =$

$\sin^{-1}x \int \dfrac{x}{\sqrt{1-x^2}}\,dx - \int\left(\dfrac{d}{dx}(\sin^{-1}x)\int \dfrac{x}{\sqrt{1-x^2}}\,dx\right)dx$

$= \sin^{-1}x \int \dfrac{x}{\sqrt{1-x^2}}\,dx - \int\left(\dfrac{1}{\sqrt{1-x^2}}\int \dfrac{x}{\sqrt{1-x^2}}\,dx\right)dx$

$= \sin^{-1}x\left(-\sqrt{1-x^2}\right) - \int\left(\dfrac{1}{\sqrt{1-x^2}}\left(-\sqrt{1-x^2}\right)\right)dx$

$= \sin^{-1}x\left(-\sqrt{1-x^2}\right) - \int(-1)\,dx$

$= x - \sin^{-1}x . \sqrt{1-x^2} + c$

12. $\dfrac{dr}{dt} = -0.02; \qquad \dfrac{dh}{dt} = 0.01$

$$V = r^2\pi h$$

$\dfrac{dV}{dt} = \pi\left(2r\dfrac{dr}{dt}\right)h + \pi r^2\dfrac{dh}{dt}$

$= \pi\left(2 \times 0.6 \times (-0.02) \times 2 + 0.36 \times 0.01\right)$

$= \pi\left(-0.048 + 0.0036\right)$

$= -0.0444\pi\left(\approx -0.14\right)$

The rate of change in the volume is -0.0444π m^3 s^{-1}.

13. $x = 2t + \tfrac{1}{2}t^2 \quad\Rightarrow\quad \dfrac{dx}{dt} = 2 + t$

$y = \tfrac{1}{3}t^3 - 3t \quad\Rightarrow\quad \dfrac{dy}{dt} = t^2 - 3$

$\dfrac{dy}{dx} = \dfrac{t^2 - 3}{2 + t}$

$\dfrac{d}{dt}\left(\dfrac{dy}{dx}\right) = \dfrac{2t(2+t) - (t^2-3)}{(2+t)^2} = \dfrac{t^2 + 4t + 3}{(2+t)^2}$

$\dfrac{d^2y}{dx^2} = \dfrac{t^2 + 4t + 3}{(2+t)^2} \times \dfrac{1}{2+t} = \dfrac{t^2 + 4t + 3}{(2+t)^3}$

Stationary points when $\dfrac{dy}{dx} = 0$, i.e.

$t^2 - 3 = 0 \Rightarrow t = \pm\sqrt{3}$

When $t = \sqrt{3}$, $\dfrac{d^2y}{dx^2} = \dfrac{3 + 4\sqrt{3} + 3}{(2 + \sqrt{3})^3} > 0$
which gives a minimum.

When $t = -\sqrt{3}$, $\dfrac{d^2y}{dx^2} = \dfrac{3 - 4\sqrt{3} + 3}{(2 - \sqrt{3})^3} < 0$
which gives a maximum.

At a point of inflexion, $\dfrac{d^2y}{dx^2} = 0$.

In this case, that means

$t^2 + 4t + 3 = (t+1)(t+3) = 0$

and this has exactly two roots.

Note that this is a slimmed-down version of the complete story of points of inflexion.

14. (a)
$\begin{vmatrix} 4 & 0 & 6 & | & 1 \\ 2 & -2 & 4 & | & -1 \\ -1 & 1 & \lambda & | & 2 \end{vmatrix}$

$\begin{vmatrix} 4 & 0 & 6 & | & 1 \\ 0 & 4 & -2 & | & 3 \\ 0 & 4 & 6+4\lambda & | & 9 \end{vmatrix}$

$\begin{vmatrix} 4 & 0 & 6 & | & 1 \\ 0 & 4 & -2 & | & 3 \\ 0 & 0 & 8+4\lambda & | & 6 \end{vmatrix}$

$z = \dfrac{6}{8 + 4\lambda} = \dfrac{3}{2(2 + \lambda)}$

$4y = 3 + 2z \Rightarrow 4y = \dfrac{18 + 6\lambda}{4 + 2\lambda}$

$\Rightarrow y = \dfrac{3\lambda + 9}{4(2 + \lambda)}$

$4x = 1 - 6z \Rightarrow 4x = \dfrac{2\lambda - 14}{4 + 2\lambda}$

$\Rightarrow x = \dfrac{\lambda - 7}{4(2 + \lambda)}$

(b) When $\lambda = -2$, the final row gives $0 = 6$ which is inconsistent.
There are no solutions.

(c) $\lambda = -2.1$; $x = 22.75$; $y = -6.75$; $z = -15$
Although the values of λ are close, the values of x, y and z are quite different. The system is **ill-conditioned** near $\lambda = -2$.

15. (a) $\dfrac{1}{(x-1)(x+2)^2} = \dfrac{A}{x-1} + \dfrac{B}{x+2} + \dfrac{C}{(x+2)^2}$

$1 = A(x+2)^2 + B(x-1)(x+2) + C(x-1)$

$x = 1 \Rightarrow A = \tfrac{1}{9}$

$x = -2 \Rightarrow C = -\tfrac{1}{3}$

$x = 0 \Rightarrow 1 = \tfrac{4}{9} - 2B + \tfrac{1}{3} \Rightarrow B = -\tfrac{1}{9}$

$\therefore \dfrac{1}{(x-1)(x+2)^2} = \dfrac{1}{9}\left(\dfrac{1}{x-1} - \dfrac{1}{x+2} - \dfrac{3}{(x+2)^2}\right)$

(b) $(x-1)\dfrac{dy}{dx} - y = \dfrac{x-1}{(x+2)^2}$

$\dfrac{dy}{dx} - \dfrac{1}{x-1}y = \dfrac{1}{(x+2)^2}$

Integrating factor: $\exp\left(\int -\dfrac{1}{x-1}dx\right)$

$= \exp\left(-\ln(x-1)\right) = (x-1)^{-1}$

$\dfrac{1}{(x-1)}\dfrac{dy}{dx} - \dfrac{1}{(x-1)^2}y = \dfrac{1}{(x-1)(x+2)^2}$

$\dfrac{d}{dx}\left(\dfrac{y}{x-1}\right) = \dfrac{1}{(x-1)(x+2)^2}$

$= \dfrac{1}{9}\left(\dfrac{1}{x-1} - \dfrac{1}{x+2} - \dfrac{3}{(x+2)^2}\right)$

$\dfrac{y}{x-1} = \dfrac{1}{9}\left(\ln|x-1| - \ln|x+2| + \dfrac{3}{x+2}\right) + c$

$y = \dfrac{x-1}{9}\left(\ln|x-1| - \ln|x+2| + \dfrac{3}{x+2}\right) + c(x-1)$

$= \dfrac{x-1}{9}\left(\ln\dfrac{|x-1|}{|x+2|} + \dfrac{3}{x+2}\right) + c(x-1)$

16. (a) For $n = 1$, the LHS $= \cos\theta + i\sin\theta$ and the RHS $= \cos\theta + i\sin\theta$. Hence the result is true for $n = 1$.

Assume the result is true for $n = k$, i.e.
$(\cos\theta + i\sin\theta)^k = \cos k\theta + i\sin k\theta$.

Now consider the case when $n = k + 1$:
$(\cos\theta + i\sin\theta)^{k+1} = (\cos\theta + i\sin\theta)^k(\cos\theta + i\sin\theta)$
$= (\cos k\theta + i\sin k\theta)(\cos\theta + i\sin\theta)$
$= (\cos k\theta\cos\theta - \sin k\theta\sin\theta) + i(\sin k\theta\cos\theta + \cos k\theta\sin\theta)$
$= \cos(k+1)\theta + i\sin(k+1)\theta$
Thus, if the result is true for $n = k$ the result is true for $n = k + 1$.
Since it is true for $n = 1$, the result is true for all $n \geq 1$.

(b) $\dfrac{(\cos\frac{\pi}{18} + i\sin\frac{\pi}{18})^{11}}{(\cos\frac{\pi}{36} + i\sin\frac{\pi}{36})^4} = \dfrac{\cos\frac{11\pi}{18} + i\sin\frac{11\pi}{18}}{\cos\frac{\pi}{9} + i\sin\frac{\pi}{9}}$

$= \dfrac{\cos\frac{11\pi}{18} + i\sin\frac{11\pi}{18}}{\cos\frac{\pi}{9} + i\sin\frac{\pi}{9}} \times \dfrac{\cos\frac{\pi}{9} - i\sin\frac{\pi}{9}}{\cos\frac{\pi}{9} - i\sin\frac{\pi}{9}}$

$= \dfrac{\cos\frac{11\pi}{18}\cos\frac{\pi}{9} + \sin\frac{11\pi}{18}\sin\frac{\pi}{9}}{\cos^2\frac{\pi}{9} + \sin^2\frac{\pi}{9}} + \text{imaginary term}$

$= \cos\left(\dfrac{11\pi}{18} - \dfrac{\pi}{9}\right) + \text{imaginary term}$

$= \cos\dfrac{\pi}{2} + \text{imaginary term}$

Thus the real part is zero as required.

---

1. $\,^4C_0(3x)^4\left(\dfrac{-2}{x^2}\right) + \,^4C_1(3x)^3\left(\dfrac{-2}{x^2}\right)^1 + \,^4C_2(3x)^2\left(\dfrac{-2}{x^2}\right)^2$

$+ \,^4C_3(3x)^1\left(\dfrac{-2}{x^2}\right)^3 + \,^4C_4(3x)^0\left(\dfrac{-2}{x^2}\right)^4$

$= 81x^4 + 4.27\,x^3 \cdot \dfrac{-2}{x^2} + 6.9x^2 \cdot \dfrac{4}{x^4} + 4.3x \cdot \dfrac{-8}{x^6} + \dfrac{16}{x^8}$

$= 81x^4 - 216\,x + \dfrac{216}{x^2} - \dfrac{96}{x^5} + \dfrac{16}{x^8}$

2. $f'(x) = e^{\cos x}(-\sin x).\sin^2 x + e^{\cos x}.2\sin x\cos x$

$= -e^{\cos x}\sin^3 x + e^{\cos x}.2\sin x\cos x$ **OR**
$= e^{\cos x}(\sin 2x - \sin^3 x)$ **OR**
$= e^{\cos x}\sin x(2\cos x - \sin^2 x)$

3. (a) $A^2 = \begin{pmatrix} 4 & p \\ -2 & 1 \end{pmatrix}\begin{pmatrix} 4 & p \\ -2 & 1 \end{pmatrix} = \begin{pmatrix} 16 - 2p & 4p + p \\ -8 - 2 & -2p + 1 \end{pmatrix}$

$= \begin{pmatrix} 16 - 2p & 5p \\ -10 & 1 - 2p \end{pmatrix}$

(b) A^2 is singular when $\det A^2 = 0$
$(16 - 2p)(1 - 2p) + 50p = 0$
$16 - 34p + 4p^2 + 50p = 0$
$4p^2 + 16p + 16 = 0$
$4(p + 2)^2 = 0$
$\mathbf{p = -2}$

OR

A^2 is singular when A is singular, [i.e. when $\det A = 0$]

$4 + 2p = 0$
$\mathbf{p = -2}$

(c) $A' = \begin{pmatrix} 4 & -2 \\ p & 1 \end{pmatrix}$

$\begin{pmatrix} x & -6 \\ 1 & 3 \end{pmatrix} = 3\begin{pmatrix} 4 & -2 \\ p & 1 \end{pmatrix}$

$\begin{pmatrix} x & -6 \\ 1 & 3 \end{pmatrix} = \begin{pmatrix} 12 & -6 \\ 3p & 3 \end{pmatrix}$ $x = 12,\ p = \dfrac{1}{3}$

4. (a) $a = \dfrac{dv}{dt}$

$= 3e^{3t} + 3e^t$

(b) $s = \int_0^{\ln 3} v\,dt = \int_0^{\ln 3}\left(e^{3t} + 2e^t\right)dt$

$= \left[\dfrac{1}{3}e^{3t} + 2e^t\right]_0^{\ln 3}$

$= \left(\dfrac{1}{3}e^{3\ln 3} + 2e^{\ln 3}\right) - \left(\dfrac{1}{3} + 2\right)$

$= \dfrac{1}{3}e^{\ln 3^3} + 2e^{\ln 3} - \dfrac{7}{3}$

$= \dfrac{1}{3} \times 27 + 2 \times 3 - \dfrac{7}{3}$

$= \dfrac{38}{3}$ or $12\dfrac{2}{3}$ or equivalent

5. $1204 = 1 \times 833 + 371$

$833 = 2 \times 371 + 91$

$371 = 4 \times 91 + 7$

$91 = 13 \times 7 \qquad$ so gcd is 7

$7 = 371 - 4 \times 91$

$= 371 - 4 (833 - 2 \times 371)$

$= 9 \times 371 - 4 \times 833$

$= 9(1204 - 1 \times 833) - 4 \times 833$

$= 9 \times 1204 - 13 \times 833$

$(a = 9, b = -13)$

6. $\dfrac{f'(x)}{f(x)}$

$= \dfrac{1}{3} \dots$

$\dots \ln \dots$

$\dots |1 + \tan 3x|$

$= \dfrac{1}{3} \ln|1 + \tan 3x| + c$

OR

$u = 1 + \tan 3x \qquad$ **OR** $\qquad u = \tan 3x$

$\dfrac{du}{dx} = 3\sec^2 3x$

$\dfrac{1}{3} du = 3\sec^2 3x \, dx$

$\int \dfrac{\frac{1}{3} du}{u} \qquad$ **OR** $\qquad \int \dfrac{\frac{1}{3} du}{1 + u} = \dots$

$= \dfrac{1}{3} \ln|u| + c \quad$ **OR** $\quad = \dfrac{1}{3} \ln|1 + u| + c$

$= \dfrac{1}{3} \ln|1 + \tan 3x| + c$

7. $\overline{z} = 1 + \sqrt{3}\, i$

$\overline{z} = 2\left(\cos \dfrac{\pi}{3} + i \sin \dfrac{\pi}{3}\right)$

$\overline{z}^2 = \left[2\left(\cos \dfrac{\pi}{3} + i \sin \dfrac{\pi}{3}\right)\right]^2 = 4\left(\cos \dfrac{2\pi}{3} + i \sin \dfrac{2\pi}{3}\right)$

OR

$\overline{z}^2 = \left(1 + \sqrt{3}\, i\right)^2 = 1 + 2\sqrt{3}\, i - 3 = -2 + 2\sqrt{3}\, i$

$\overline{z}^2 = -2 + 2\sqrt{3}\, i = r(\cos\theta + i \sin\theta)$

$r = 4 . \theta = \dfrac{2\pi}{3} . \overline{z}^2 = 4\left(\cos \dfrac{2\pi}{3} + i \sin \dfrac{2\pi}{3}\right)$

8. $\left[x^2 . \dfrac{1}{3} \sin 3x\right] = \int \dfrac{2}{3} x \sin 3x \, dx$

$= \left[\dfrac{1}{3} x^2 \sin 3x\right] - \left[-\dfrac{2}{9} x\cos 3x - \int -\dfrac{2}{9} \cos 3x \, dx\right]$

$= \dfrac{1}{3} x^2 \sin 3x + \dfrac{2}{9} x\cos 3x - \dfrac{2}{27} \sin 3x + c$

9. $\displaystyle\sum_{r=1}^{n} \left(4r^3 + 3r^2 + r\right) = n(n+1)^3.$

For $n = 1$

L.H.S. $\qquad\qquad\qquad$ R.H.S.

$\displaystyle\sum_{r=1}^{n} \left(4r^3 + 3r^2 + r\right) \qquad n(n+1)^3.$

$= 4 + 3 + 1 = 8 \qquad\qquad = 1 \times 2^3 = 8$

$\Rightarrow$ true for $n = 1$

Assume true for $n = k$,

$\displaystyle\sum_{r=1}^{k} \left(4r^3 + 3r^2 + r\right) = k(k+1)^3$

Consider $n = k + 1$,

$\displaystyle\sum_{r=1}^{k+1} \left(4r^3 + 3r^2 + r\right)$

$= \displaystyle\sum_{r=1}^{k} \left(4r^3 + 3r^2 + r\right) + 4(k+1)^3 + 3(k+1)^2 + (k+1)$

$= k(k+1)^3 + 4(k+1)^3 + 3(k+1)^2 + (k+1)$

$= (k+1)\left[k(k+1)^2 + 4(k+1)^2 + 3(k+1) + 1\right]$

$= (k+1)\left[k(k^2 + 2k + 1) + 4(k^2 + 2k + 1) + 3(k+1) + 1\right]$

$= (k+1)\left[k^3 + 2k^2 + k + 4k^2 + 8k + 4 + 3k + 3 + 1\right]$

$= (k+1)\left(k^3 + 6k^2 + 12k + 8\right)$

$= (k+1)(k+2)^3$

$= (k+1)((k+1)+1)^3$

Hence, if true for $n = k$, then true for $n = k + 1$, but since true for $n = 1$, then by induction true for all positive integers n.

10. (a) Circle...

$\qquad\qquad$ ...centre $(0, -1)$ [or $-i$], radius 1

$\qquad$ **OR** $\qquad z + i = x + iy + i = x + i(y + 1)$

$\qquad\qquad\qquad |x + (y + 1)i|^2 = 1$

$x^2 + (y + 1)^2 = 1$

$\qquad\qquad$ Circle centre $(0, -1)$, radius 1

$\qquad$ **OR**

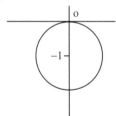

(b) Set of points equidistant from $(1, 0)$ and $(-5, 0)$

Straight line...

$$...x = -2$$

OR $\quad |z-1|^2 \qquad = |z+5|^2$

$$|(x-1) + iy|^2 = |(x+5) + iy|^2$$
$$(x-1)^2 + y^2 \qquad = (x+5)^2 + y^2$$
$$-2x + 1 \qquad = 10x + 25$$
$$-24 \qquad = 12x$$
$$x \qquad = -2$$

which is a **straight line**

OR

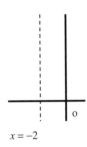

$$x = -2$$

11. $2x + 4x\dfrac{dy}{dx} + 4y ...$

$$... + 2y\dfrac{dy}{dx} = 0 \qquad (\varDelta)$$

$$2(-2) + 4(-2)\dfrac{dy}{dx} + 4(3) + 2(3)\dfrac{dy}{dx} = 0 \quad \therefore \dfrac{dy}{dx} = 4$$

OR $\dfrac{dy}{dx} = -\dfrac{2x+4y}{4x+2y} = -\dfrac{x+2y}{2x+y} \quad (\dagger) \ \therefore \dfrac{dy}{dx} = 4$

Differentiating (Δ): $2 + 4x\dfrac{d^2y}{dx^2} + 4\dfrac{dy}{dx} + 4\dfrac{dy}{dx} ...$

$$... + 2y\dfrac{d^2y}{dx^2} + 2\left(\dfrac{dy}{dx}\right)^2 = 0$$

$$\therefore \ 2 + 4(-2)\dfrac{d^2y}{dx^2} + 8(4) + 2(3)\dfrac{d^2y}{dx^2} + 2(4)^2 = 0$$

$$\therefore \ \dfrac{d^2y}{dx^2} = 33$$

OR Differentiating ($\dagger$):

$$\dfrac{d^2y}{dx^2} = -\dfrac{(2x+y)\left(1 + 2\dfrac{dy}{dx}\right) - (x+2y)\left(2 + \dfrac{dy}{dx}\right)}{(2x+y)^2}$$

$$\dfrac{d^2y}{dx^2} = -\dfrac{(2(-2)+3)(1 + 2(4)) - ((-2)+2(3))(2+4)}{(2(-2)+3)^2} = 33$$

12. A Suppose $n = 9m$ for some natural number [positive integer], m

Then $n^2 = 81m^2 = 9(9m^2)$

Hence n^2 is a multiple of 9, so **A** is **true**.

B **False.** Use any valid counterexample:
$n = 3, 6, 12 , 15$ or 21 etc

13. (a)

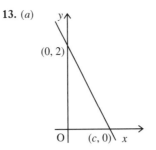

(b) $y = f(x) - c$ is odd. $\therefore \boldsymbol{k = -c}$

(c)

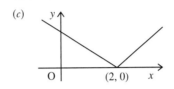

$y = |f(x+2)|$ is even $\therefore \boldsymbol{h = 2}$

14. $m^2 - 6m + 9 = 0$
$$(m-3)^2 = 0$$
$$m = 3$$

C.F. $\quad y = Ae^{3x} + Bxe^{3x}$

P.I. Try $y = Cx^2e^{3x}$ (since e^{3x} and xe^{3x} in c.f.)

$$\dfrac{dy}{dx} = 2Cxe^{3x} + 3Cx^2e^{3x}$$

$$\dfrac{d^2y}{dx^2} = 2Ce^{3x} + 6Cxe^{3x} + 6Cxe^{3x} + 9Cx^2e^{3x}$$

$$= 2Ce^{3x} + 6Cxe^{3x} + 6Cxe^{3x} + 9Cx^2e^{3x}$$
$$= -6(2Cxe^{3x} + 3Cx^2e^{3x}) + 9Cx^2e^{3x} = 4e^{3x}$$
$$2Ce^{3x} = 4e^{3x} \Rightarrow C = 2$$

G.S. $\quad y = Ae^{3x} + Bxe^{3x} + 2x^2e^{3x}$

$$\dfrac{dy}{dx} = 3Ae^{3x} + Be^{3x} + 3Bxe^{3x} + 4xe^{3x} + 6x^2e^{3x}$$

When $x = 0$, $y = 1$ $\qquad A = 1$

$\dfrac{dy}{dx} = -1$ $\qquad\qquad -1 = 3 + B \Rightarrow B = -4$

P.S. $\quad \boldsymbol{y = e^{3x} - 4xe^{3x} + 2x^2e^{3x}}$

15. (*Any two*)

(a) $\overrightarrow{AB} = \begin{pmatrix} 1 \\ 1 \\ 0 \end{pmatrix}$ $\qquad \overrightarrow{AC} = \begin{pmatrix} 0 \\ 1 \\ 2 \end{pmatrix}$ $\qquad$ **OR** $\qquad \overrightarrow{BC} = \begin{pmatrix} -1 \\ 0 \\ 2 \end{pmatrix}$

$$\overrightarrow{AB} \times \overrightarrow{AC} = \begin{vmatrix} i & j & k \\ 1 & 1 & 0 \\ 0 & 1 & 2 \end{vmatrix}$$ or equivalent

$$= 2\mathbf{i} - 2\mathbf{j} + \mathbf{k}$$

$$2x - 2y + z = 2 \times 0 - 2 \times (-1) + 1 \times 3 = 5$$

$$\pi_1 : \boldsymbol{2x - 2y + z = 5}$$

OR $\quad \mathbf{r} = \begin{pmatrix} 0 \\ -1 \\ 3 \end{pmatrix} + \lambda\begin{pmatrix} 1 \\ 1 \\ 0 \end{pmatrix} + \mu\begin{pmatrix} 0 \\ 1 \\ 2 \end{pmatrix}$ or equivalent

(b) $0 \times 0 + (-1) \times (-1) + 1 \times 3 = 4$

$$\pi_2 : \boldsymbol{-y + z = 4}$$

(c) Normal vectors:

$$n_1 = \begin{pmatrix} 2 \\ -2 \\ 1 \end{pmatrix} \text{ and } n_2 = \begin{pmatrix} 0 \\ -1 \\ 1 \end{pmatrix}, \; |n_1| = \sqrt{9} = 3, \; |n_2| = \sqrt{2}$$

cos (angle between normals) =

$$\frac{n_1 n_2}{|n_1||n_2|} = \frac{2 \times 0 - 2 \times -1 + 1 \times 1}{3\sqrt{2}} = \frac{3}{3\sqrt{2}} = \frac{1}{\sqrt{2}}$$

Angle $= 45°$

acute angle between planes is **45°** $\left(\text{or } \dfrac{\pi}{4}\right)$.

OR

$n_1 = 2i - 2j + k$, so $|2i - 2j + k| = 3$ and $|-j + k| = \sqrt{2}$

$$3 = |n_1| . |n_2| . \cos\Theta = 3\sqrt{2} . \cos\Theta$$

$$\cos\Theta = \frac{1}{\sqrt{2}} \text{ so } \Theta = \frac{\pi}{4} \text{ (or } 45°)$$

16. $\dfrac{dP}{dt} = P(1000 - P)$

So $\displaystyle\int \frac{dP}{P(1000 - P)} = \int dt$

$$\frac{1}{P(1000 - P)} = \frac{A}{P} + \frac{B}{1000 - P}$$

$A = \dfrac{1}{1000}, B = \dfrac{1}{1000}$

$$\frac{1}{1000}\int \left(\frac{1}{P} + \frac{1}{1000 - P}\right) dP = \int dt$$

$\ln P - \ln(1000 - P) = 1000t + c$

$\ln \dfrac{P}{1000 - P} = 1000t + c$

$\dfrac{P}{1000 - P} = Ke^{1000t} \; \left(\text{where } K = e^c\right)$

$P = 1000Ke^{1000t} - PKe^{1000t}$,

$P + PKe^{1000t} = 1000Ke^{1000t}$,

$P = \dfrac{1000Ke^{1000t}}{1 + Ke^{1000t}}$

$\quad = \dfrac{1000K}{e^{-1000t} + K} \quad \left(\text{or } \dfrac{1000e^c}{e^{-1000t} + e^c}\right)$

Since $P(0) = 200$, $\quad 200 = \dfrac{1000K}{1 + K} \quad 200 + 200K = 1000K$

$\therefore K = \dfrac{1}{4}$ (or 0·25)

Require $\quad 900 = \dfrac{1000 \times 0·25}{0·25 + e^{-1000t}}$

$225 + 900e^{-1000t} = 250$

$e^{1000t} = 36$

$1000t = \ln 36$

$t = \dfrac{1}{1000} \ln 36$

[or 0·003584 (4sf)] **OR** $\quad t = -\dfrac{1}{1000}\ln\left(\dfrac{1}{36}\right)$

17. $1 + x + x^2 + x^3 + \ldots = \dfrac{1}{1 - x}$

$1 - x + x^2 - x^3 + \ldots = \dfrac{1}{1 + x}$

Integrating the first of these gives:

$x + \dfrac{1}{2}x^2 + \dfrac{1}{3}x^3 + \dfrac{1}{4}x^4 + \dfrac{1}{5}x^5 + \ldots = -\ln(1 - x) + c$

Putting $x = 0$ gives $c = 0$.

Similarly, $x - \dfrac{1}{2}x^2 + \dfrac{1}{3}x^3 + \dfrac{1}{4}x^4 \ldots = \ln(1 + x) + c$

Putting $x = 0$ gives $c = 0$ again!

Adding together gives:

$2\left(x + \dfrac{1}{3}x^3 + \dfrac{1}{5}x^5 + \ldots\right) = \ln(1 + x) - \ln(1 - x)$

$\left[= \ln\dfrac{1 + x}{1 - x}\right]$ as required.

OR $\quad 2 + 2x^2 + 2x^4 + \ldots = \dfrac{1}{1 + x} + \dfrac{1}{1 - x}$

$\therefore 2x + \dfrac{2}{3}x^3 + \dfrac{2}{5}x^5 + \ldots$

$\qquad = \ln(1 + x) \ldots$

$\qquad \ldots -\ln(1 - x) + c$

Putting $x = 0$ gives $c = 0$.

$\left[= \ln\dfrac{1 + x}{1 - x}\right]$ as required.

OR

$f(x) = \ln\left(\frac{1+x}{1-x}\right) = \ln(1 + x) - \ln(1 - x)$	$f(0) = 0$
$f'(x) = 2(1 - x^2)^{-1}$ or equivalent	$f'(0) = 2$
$f''(x) = 4x(1 - x^2)^{-2}$	$f''(0) = 0$
$f'''(x) = 16x^2(1 - x^2)^{-3} + 4(1 - x^2)^{-2}$	$f'''(0) = 4$
$f^{IV}(x) = 96x^3(1 - x^2)^{-4} + 48x(1 - x^2)^{-3}$	$f^{IV}(0) = 0$
$f^{V}(x) = 768x^4(1 - x^2)^{-5} + $ $576x^2(1 - x^2)^{-4} + 48(1 - x^2)^{-3}$	$f^{V}(0) = 48$

$\therefore f(x) = 0 + 2.1x + 0x^2 + \dfrac{4}{3!}x^3 + 0x^4 + \dfrac{48}{5!}x^5$

$\quad = 2x + \dfrac{2}{3}x^3 + \dfrac{2}{5}x^5 + \ldots$

so $f(x) = \ln\left(\dfrac{1 + x}{1 - x}\right) = 2\left(x + \dfrac{x^3}{3} + \dfrac{x^5}{5} + \ldots\right)$ as required.

Now choose x such that $\dfrac{1 + x}{1 - x} = 2$,

i.e. $1 + x = 2 - 2x$, so $x = \dfrac{1}{3}$

So $\ln 2 = 2\left(\dfrac{1}{3} + \dfrac{1}{81} + \dfrac{1}{1215} + \dfrac{1}{15309} + \ldots\right)$

ADVANCED HIGHER MATHEMATICS 2014

1. (a) $f'(x) = \dfrac{(x^2+1).2x - (x^2+1).2x}{(x^2+1)^2}$

$= \dfrac{2x^3 + 2x - 2x^3 + 2x}{(x^2+1)^2}$

$= \dfrac{4x}{(x^2+1)^2}$

OR

$f(x) = 1 - \dfrac{2}{x^2+1}$

$f'(x) = -1(-2)(x^2+1)^{-2}\dots$

$\dots \times 2x$

$\therefore f'(x) = 4x^2(x^2+1)^{-2}$

$= \dfrac{4x}{(x^2+1)^2}$

(b) $= \dfrac{6x}{1+(3x^2)^2}$

$= \dfrac{6x}{1+9x^4}$

OR

$\tan y = 3x^2$

$\sec^2 y . \dfrac{dy}{dx} = 6x$

$\dfrac{dy}{dx} = \dfrac{6x}{\sec^2 y} = \dfrac{6x}{\sec^2(\tan^{-1}(3x^2))}$

2.
$= \dbinom{10}{r}\left(\dfrac{2}{x}\right)^r \left(\dfrac{1}{4x^2}\right)^{10-r}$ **OR** $\dbinom{10}{r}\left(\dfrac{2}{x}\right)^{10-r}\left(\dfrac{1}{4x^2}\right)^r$

$= \dbinom{10}{r}\dfrac{2^r}{x^r (4x^2)^{10-r}}$ **OR** $\dbinom{10}{r}\dfrac{2^{10-r}}{x^{10-r}(4x^2)^r}$

$= \dbinom{10}{r}\dfrac{2^r}{x^r 4^{10-r}(x^2)^{10-r}}$ **OR** $\dbinom{10}{r}\dfrac{2^{10-r}}{x^{10-r}4^r(x^2)^r}$

$= \dbinom{10}{r}\dfrac{2^r}{x^r 2^{20-2r}x^{20-2r}}$ **OR** $\dbinom{10}{r}\dfrac{2^{10-r}}{x^{10-r}2^{2r}x^{2r}}$

$= \dbinom{10}{r}\dfrac{2^{3r-20}}{x^{20-r}}$ **OR** $\dbinom{10}{r}\dfrac{2^{10-3r}}{x^{10+r}}$

$= \dbinom{10}{r}2^{3r-20}x^{r-20}$ **OR** $\dbinom{10}{r}2^{10-3r}x^{-r-10}$

For term in x^{-13} : $r=7$ **OR** $r=3$

ie $= \dbinom{10}{7}2^{3\times7-20}x^{7-20}$ **OR** $\dbinom{10}{3}2^{10-3\times3}x^{-3-10}$

$= 240x^{-13}$ **OR** $\dfrac{240}{x^{13}}$

3.

1	1	1	2	R_1
4	3	$-\lambda$	4	$4R_1 - 1R_2$
5	6	8	11	$R_3 - 5R_1$

1	1	1	2	R_1
0	1	$4+\lambda$	4	R_2
0	1	3	1	$R_2 - R_3$

1	1	1	2
0	1	$4+\lambda$	4
0	0	$1+\lambda$	3

$(1+\lambda)z = 3$

$z = \dfrac{3}{1+\lambda}$

$\lambda \neq -1$

$z = 1, y = -2, x = 3.$

4. $\dfrac{dx}{dt} = \dfrac{2t}{1+t^2}$

$\dfrac{dy}{dx} = \dfrac{4t}{1+2t^2}$

$\dfrac{dy}{dx} = \dfrac{4t(1+t^2)}{2t(1+2t^2)}$

$\dfrac{dy}{dx} = \dfrac{2(1+t^2)}{(1+2t^2)}$ **OR** $\dfrac{dy}{dx} = \dfrac{2+2t^2}{1+2t^2}$

5.
$\begin{vmatrix} i & j & k \\ 2 & 1 & 3 \\ 1 & 4 & -1 \end{vmatrix}$

$= -13i + 5j + 7k$

$u.(v \times w) = \begin{pmatrix} 5 \\ 13 \\ 0 \end{pmatrix}.\begin{pmatrix} -13 \\ 5 \\ 7 \end{pmatrix} = 0$

u lies in the same plane as the one containing both v and w.
OR u is parrallel to the plane containing v and w.
OR u is perpendicular to the normal of v and w.
OR All **4** points lie in same plane.
OR u is perpendicular to $v \times w$.
OR Volume of parallelepiped is zero.
OR u, v and w are coplanar/linearly dependent.

OR $u.(v \times w) = \begin{vmatrix} 5 & 13 & 0 \\ 2 & 1 & 3 \\ 1 & 4 & -1 \end{vmatrix}$

$= 5\begin{vmatrix} 1 & 3 \\ 4 & -1 \end{vmatrix} -13\begin{vmatrix} 2 & 3 \\ 1 & -1 \end{vmatrix} + 0\begin{vmatrix} 2 & 1 \\ 1 & 4 \end{vmatrix}$

$= 0$

6. $y = \ln(x^3 \cos^2 x)$

$y = \ln(x^3) + \ln(\cos^2 x)$

$\dfrac{dy}{dx} = \dfrac{3}{x} - \dfrac{2 \sin x}{\cos x}$

$\dfrac{dy}{dx} = \dfrac{3}{x} - 2 \tan x$

a = 3, b = -2.

OR

$y = \ln(x^3 \cos^2 x)$

$\dfrac{dy}{dx} = \dfrac{3x^2 \cos^2 x - 2x^3 \sin x \cos x}{x^3 \cos^2 x}$

$\dfrac{dy}{dx} = \dfrac{3}{x} - 2 \tan x$

a = 3, b = -2.

OR

$e^y \dfrac{dy}{dx} = 3x^2 \cos^2 x - 2x^3 \sin x \cos x$

$\dfrac{dy}{dx} = \dfrac{3x^2 \cos^2 x - 2x^3 \sin x \cos x}{x^3 \cos^2 x}$

$\dfrac{dy}{dx} = \dfrac{3}{x} - 2 \tan x$

a = 3, b = -2.

7. For $n = 1$ RHS $= \begin{pmatrix} 2^1 & a(2^1 - 1) \\ 0 & 1 \end{pmatrix}$

$= \begin{pmatrix} 2 & a \\ 0 & 1 \end{pmatrix}$

$= A$

LHS $= A^1 = A = $ RHS

Assume true for $n = k$,

$A^k = \begin{pmatrix} 2^k & a(2^k - 1) \\ 0 & 1 \end{pmatrix}$

Consider $n = k + 1$,

$A^{k+1} = A^k A^1$ **[OR** $A^1 A^k$**]**

$= \begin{pmatrix} 2^k & a(2^k - 1) \\ 0 & 1 \end{pmatrix} \begin{pmatrix} 2 & a \\ 0 & 1 \end{pmatrix}$

$= \begin{pmatrix} 2^k.2 & 2^k.a + a(2^k - 1) \\ 0 & 1 \end{pmatrix}$

$= \begin{pmatrix} 2^{k+1} & 2^k.a + 2^k.a - a \\ 0 & 1 \end{pmatrix}$

$= \begin{pmatrix} 2^{k+1} & a(2^k + 2^k - 1) \\ 0 & 1 \end{pmatrix}$

$= \begin{pmatrix} 2^{k+1} & a(2^{k+1} - 1) \\ 0 & 1 \end{pmatrix}$

Hence, if true for $n = k$, then true for $n = k + 1$, but since true for $n = 1$, then by induction true for all positive integers n.

8. $4m^2 - 4m + 1 = 0$

$(2m - 1)^2 = 0$

$m = \dfrac{1}{2}$

C.F./G.S. $y = Ae^{\frac{1}{2}x} + Bxe^{\frac{1}{2}x}$

$y = 4$ when $x = 0$ gives $4 = A.1 + 0$, so $A = 4$

$\dfrac{dy}{dx} = \dfrac{1}{2} Ae^{\frac{1}{2}x} + Be^{\frac{1}{2}x} + \dfrac{1}{2} Bxe^{\frac{1}{2}x}$

$\dfrac{dy}{dx} = 3$ when x = 0 gives

$3 = \dfrac{1}{2} Ae^0 + Bxe^0 + \dfrac{1}{2} B.0.e^0$

$3 = \dfrac{1}{2}.4 + B$,   so $B = 1$

So P.S. is: $y = 4e^{\frac{1}{2}x} + xe^{\frac{1}{2}x}$

9. $\cos x = 1 - \dfrac{x^2}{2!} + \dfrac{x^4}{4!} \dots$

$\cos 3x = 1 - \dfrac{(3x)^2}{2!} + \dfrac{(3x)^4}{4!} \dots$

$= 1 - \dfrac{9x^2}{2} + \dfrac{81x^4}{24} \dots$

$= 1 - \dfrac{9x^2}{2} + \dfrac{27x^4}{8} \dots$

$e^{2x} = 1 + 2x + \dfrac{(2x)^2}{2!} + \dfrac{(2x)^3}{3!} + \dots$

$= 1 + 2x + 2x^2 + \dfrac{4x^3}{3} + \dots$

$e^{2x} \cos 3x = \left(1 - \dfrac{9x^2}{2} + \dfrac{27x^4}{8} \dots\right)\left(1 + 2x + 2x^2 + \dfrac{4x^3}{3} \dots\right)$

$= 1 + 2x + 2x^2 + \dfrac{4x^3}{3} - \dfrac{9x^2}{2} - \dfrac{18x^3}{2} \dots$

$= 1 + 2x - \dfrac{5x^2}{2} - \dfrac{23x^3}{3}$

10. $(x - 1)^2 + y^2 = 4$

$V = \pi \displaystyle\int_0^3 y^2 \, dx$

$= \pi \displaystyle\int_0^3 \left(4 - (x - 1)^2\right) dx$

$= \pi \left[4x - \dfrac{1}{3}(x - 1)^3\right]_0^3$

$= 9\pi$ **units**^{3}

11. (a)

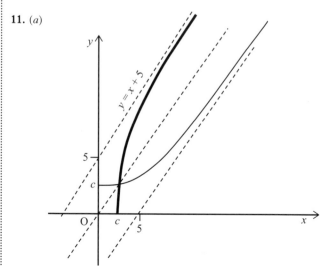

(b) $y = x - 3$

(c) From the diagram, the two **curves/graphs** intersect

OR

$y = f(x)$ intersects $y = x$

OR

$y = f^{-1}(x)$ intersects $y = x$

OR

$f^{-1}(x) = f(x)$

So $x = f(f(x))$

12.
$$x = \tan\theta$$

$$\frac{dx}{d\theta} = \sec^2\theta$$

$$dx = \sec^2\theta\, d\theta$$

$x = 1$ and $x = 0$ become $\theta = \dfrac{\pi}{4}$ and $\theta = 0$

$$\int_0^{\frac{\pi}{4}} \frac{\sec^2\theta\, d\theta}{\left(1 + \tan^2\theta\right)^{\frac{3}{2}}}$$

$$\int_0^{\frac{\pi}{4}} \frac{\sec^2\theta\, d\theta}{\left(\sec^2\theta\right)^{\frac{3}{2}}}$$

$$\int_0^{\frac{\pi}{4}} \cos\theta\, d\theta$$

$$= \left[\sin\theta\right]_0^{\frac{\pi}{4}} = \frac{1}{\sqrt{2}}$$

13. $\dfrac{dF}{dx} = 0 + e^x(\cos x + \sin x) + e^x\left(\sin x - \cos x - \sqrt{2}\right)$

$\qquad = e^x\left(2\sin x - \sqrt{2}\right)$

For S.P.s, $\dfrac{dF}{dx} = 0$, so $\qquad e^x\left(2\sin x - \sqrt{2}\right) = 0$

Then: $2\sin x = \sqrt{2}$ and so $\sin x = \dfrac{1}{\sqrt{2}}$

Hence $x = \dfrac{\pi}{4},\ \dfrac{3\pi}{4}$

Leading to values of $F = 11\cdot9,\ 15$

$$40 \le s \le 120 \text{ so } 0 \le x \le \pi$$

$$x = 0,\ F = 12\cdot6;\quad x = \pi,\ F = 5\cdot4$$

Greatest efficiency 15 km/litre at 100 km/h.

Least efficiency 5·4 km/litre at 120 km/h.

14. (a) $\qquad 1 + r + r^2 + r^3 + \ldots = \dfrac{1}{1 - r}$

$$\frac{1}{2 - 3r} = \frac{1}{2\left(1 - \dfrac{3r}{2}\right)} \quad \textbf{OR} \quad \frac{1}{1 - (3r - 1)} \quad \textbf{OR} \quad \frac{\frac{1}{2}}{1 - \dfrac{3}{2}r}$$

$$= \frac{1}{2}\left(\frac{1}{1 - \dfrac{3r}{2}}\right) = \frac{1}{2}\left(1 + \frac{3r}{2} + \left(\frac{3r}{2}\right)^2 + \ldots\right)$$

$$= \frac{1}{2}\left(1 + \frac{3r}{2} + \frac{9r^2}{4} + \ldots\right)$$

$$\left|\frac{3r}{2}\right| < 1, \quad \therefore |r| < \frac{2}{3}$$

(b)
$$\frac{1}{3r^2 - 5r + 2} = \frac{A}{(3r - 2)} = \frac{B}{(r - 1)}$$

$$\therefore A(r - 1) + B(3r - 2) \equiv 1; \qquad B = 1$$

$$A = -3$$

$$\frac{1}{3r^2 - 5r + 2} = \frac{-3}{(3r - 2)} = \frac{1}{(r - 1)}$$

$$= \frac{3}{(2 - 3r)} = \frac{1}{(1 - r)}$$

$$= 3\left(\frac{1}{2}\left(1 + \frac{3r}{2} + \frac{9r^2}{4} + \ldots\right)\right) - \left(1 + r + r^2 + \ldots\right)$$

$$= \frac{1}{2} + \frac{5r}{4} + \frac{19r^2}{8} \ldots$$

$$\left|\frac{3r}{2}\right| < 1 \text{ and } |r| < 1, \text{ so } |r| < \frac{2}{3}$$

OR

$$f(x) = (3r^2 - 5r + 2)^{-1} \qquad\qquad f(0) = \frac{1}{2}$$

$$f'(x) = -(3r^2 - 5r + 2)^{-2}(6r - 5) \qquad f'(0) = \frac{5}{4}$$

$$f''(x) = -6(3r^2 - 5r + 2)^{-2} + 2(3r^2 - 5r + 2)^{-3}(6r - 5)^2$$

$$f''(0) = \frac{19}{4}$$

$$\therefore f(x) = \frac{1}{2} + \frac{5r}{4} + \frac{19r^2}{8} \ldots$$

15. (a) $\displaystyle\int e^x \cos x\, dx = e^x \cos x - \int e^x\left(\frac{d}{dx}(\cos x)\right) dx$

$\qquad\qquad = e^x \cos x + \displaystyle\int e^x \sin x\, dx$

$\qquad\qquad = e^x \cos x + e^x \sin x - \displaystyle\int e^x \cos x\, dx$

$\therefore 2\displaystyle\int e^x \cos x\, dx = e^x \sin x + e^x \cos x + c$

$\therefore \displaystyle\int e^x \cos x\, dx = \frac{1}{2} e^x(\sin x + \cos x) + c$

OR

$\displaystyle\int e^x \cos x\, dx = e^x \sin x - \int e^x\left(\int \cos x\, dx\right) dx$

$\qquad\qquad = e^x \sin x - \displaystyle\int e^x \sin x\, dx$

$\qquad\qquad = e^x \sin x - \left(-e^x \cos x - \displaystyle\int -e^x \cos x\, dx\right)$

$\qquad\qquad = e^x \sin x + e^x \cos x - \displaystyle\int e^x \cos x\, dx$

$\therefore 2\displaystyle\int e^x \cos x\, dx = e^x \sin x + e^x \cos x + c$

$\therefore \displaystyle\int e^x \cos x\, dx = \frac{1}{2} e^x(\sin x + \cos x) + c$

(b) $I_n = e^x \cos nx - \displaystyle\int e^x(-n \sin nx)\, dx$

$\qquad = e^x \cos nx + \displaystyle\int e^x n \sin nx\, dx$

$\qquad = e^x \cos nx + e^x n \sin nx - \displaystyle\int n^2 e^x \cos nx\, dx$

$\left(1 + n^2\right) I_n = n e^x \sin nx + e^x \cos nx + c$

$I_n = \left(\dfrac{e^x}{1 + n^2}\right)(n \sin nx + \cos nx) + c$

(c) $I_8 = \left[\left(\dfrac{e^x}{8^2 + 1}\right)(8 \sin 8x + \cos 8x)\right]_0^{\frac{\pi}{2}}$

$\qquad = \dfrac{1}{65}\left(e^{\frac{\pi}{2}} - 1\right)$

16. (a) $-1 = r (\cos \theta + i \sin \theta) = 1 (\cos \pi + i \sin \pi)$

$$\therefore z = \cos \left(\frac{\pi}{4} + \frac{2\pi k}{4} \right) + i \sin \left(\frac{\pi}{4} + \frac{2\pi k}{4} \right) ; k = 0, 1, 2, 3$$

$$0 = \pm \frac{\pi}{4}, \pm \frac{3\pi}{4}$$

$$z = \cos \left(\frac{\pi}{4} \right) + i \sin \left(\frac{\pi}{4} \right), \cos \left(\frac{3\pi}{4} \right) + i \sin \left(\frac{3\pi}{4} \right),$$

$$\cos \left(-\frac{\pi}{4} \right) + i \sin \left(-\frac{\pi}{4} \right), \cos \left(-\frac{3\pi}{4} \right) + i \sin \left(-\frac{3\pi}{4} \right)$$

$$z = \cos \left(\frac{\pi}{4} \right) \pm i \sin \left(\frac{\pi}{4} \right), \cos \left(\frac{3\pi}{4} \right) \pm i \sin \left(\frac{3\pi}{4} \right)$$

(b) $z = \pm i, \pm 1$

(c)

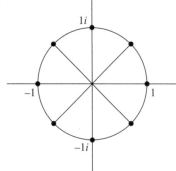

(d) $z^8 - 1 = (z^4 + 1)(z^4 - 1)$

Then the solutions to $z^4 + 1 = 0$ and $z^4 - 1 = 0$
are also the solutions to $z^8 - 1 = 0$.

(e) Observe that $z^6 + z^4 + z^2 + 1 = (z^2 + 1)(z^4 + 1)$

OR

$$z^8 - 1 = (z^4 + 1)(z^2 + 1)(z^2 - 1)$$

$\therefore$ Six solutions are those above except $z = \pm 1$